T0176259

DISCOVER SIGNAL PROCESSING

DISCOVER SIGNAL PROCESSING

AN INTERACTIVE GUIDE FOR ENGINEERS

Simon Braun

Israel Institute of Technology, Israel

John Wiley & Sons, Ltd

Copyright © 2008   John Wiley & Sons Ltd, The Atrium, Southern Gate, Chichester,
                       West Sussex PO19 8SQ, England

                       Telephone (+44) 1243 779777

Email (for orders and customer service enquiries): cs-books@wiley.co.uk
Visit our Home Page on www.wileyeurope.com or www.wiley.com

All Rights Reserved. No part of this publication may be reproduced, stored in a retrieval system or transmitted in
any form or by any means, electronic, mechanical, photocopying, recording, scanning or otherwise, except under
the terms of the Copyright, Designs and Patents Act 1988 or under the terms of a licence issued by the Copyright
Licensing Agency Ltd, 90 Tottenham Court Road, London W1T 4LP, UK, without the permission in writing of
the Publisher. Requests to the Publisher should be addressed to the Permissions Department, John Wiley & Sons
Ltd, The Atrium, Southern Gate, Chichester, West Sussex PO19 8SQ, England, or emailed to permreq@wiley.
co.uk, or faxed to (+44) 1243 770620.

This publication is designed to provide accurate and authoritative information in regard to the subject
matter covered. It is sold on the understanding that the Publisher is not engaged in rendering professional
services. If professional advice or other expert assistance is required, the services of a competent professional
should be sought.

Other Wiley Editorial Offices

John Wiley & Sons Inc., 111 River Street, Hoboken, NJ 07030, USA

Jossey-Bass, 989 Market Street, San Francisco, CA 94103-1741, USA

Wiley-VCH Verlag GmbH, Boschstr. 12, D-69469 Weinheim, Germany

John Wiley & Sons Australia Ltd, 42 McDougall Street, Milton, Queensland 4064, Australia

John Wiley & Sons (Asia) Pte Ltd, 2 Clementi Loop #02-01, Jin Xing Distripark, Singapore 129809

John Wiley & Sons Canada Ltd, 6045 Freemont Blvd, Mississauga, ONT, L5R 4J3

Wiley also publishes its books in a variety of electronic formats. Some content that appears in print may not be
available in electronic books.

British Library Cataloguing in Publication Data

A catalogue record for this book is available from the British Library

ISBN 978-0470-51970-7

Typeset in 9/11 pt Times by Thomson Digital, New Delhi, India
Printed and bound in Great Britain by Antony Rowe Ltd, Chippenham, Wiltshire
This book is printed on acid-free paper responsibly manufactured from sustainable forestry in which at least two
trees are planted for each one used for paper production.

Contents

Preface

Signal processing is now a multidisciplinary topic, used by researchers and practitioners in diverse fields, including science, engineering, medicine, finances, behavioral sciences and others.

Modern software libraries that include dedicated languages and packages are now readily available, dramatically simplifying the development and application of signal processing techniques. The ease of application implies however that an understanding of the various techniques, often also at an intuitive level, is imperative. For both students and first time users or even practitioners with some experience, it is important to be able to choose an appropriate processing technique, and furthermore be aware of the errors involved as well as their control. These aspects are usually more difficult to master than the straightforward task of applying specific signal processing methods. It is specifically towards these objectives that the present text is oriented. As an example we may consider the topic of digital filters. This text is based on the assumption that existing software packages minimize the effort involved in designing such filters. It is however the problem of choosing the necessary filter types, their parameters, and the effect of that choice on the filtering results that is often more difficult to address. For example, the topic of actually designing filters, extensively covered in many texts on signal processing, is kept to an absolute minimum. Here, it is the *use* of the filters that is deemed to be of interest. Another example would be spectral analysis, where the aspects of engineering units and control of errors are emphasized.

The rationale is that of "learning by doing": what is discovered by actually attempting/exercising a task is better remembered and understood. One uniqueness of this book is thus an extensive set of exercises (approximately 60), covering major aspects of signal processing tools, geared at performing or testing specific tasks. They cover major topics, and address principles as well as advanced aspects of importance to the user.

Matlab is used as a platform, utilizing its extensive GUI capabilities, and must be available in order to run the exercises. These are performed via graphical elements only, and the reader is not distracted by any necessary programming. The interface is friendly, intuitive, and includes help via an instruction menu. Some exercises are relatively simple ones, intended to recall or practice simple concepts. Others are more complex, geared to develop a real understanding of issues involved. A few could be completed in minutes, some might necessitate more than one lengthy session.

Intended Audience

The minimum necessary background is that of linear dynamic systems, continuous and discrete.

The material is geared to:
  Engineering students (graduate/senior) – mechanical, aeronautical, civil
  Engineering students (EE and computer) as advanced material, complementing classic signal processing topics
  Practicing researchers and engineers

Industrial continuing education courses for electronic, mechanical, aeronautical, civil and computer engineers.

Book Organization (see figure)

The book is divided into 2 parts, each one covering the same 14 different topics. The exercises are presented in part A, the corresponding theoretical background in part B.

The organization of part A is as following: for each topic, an overview is presented, followed first by the exercises themselves, and finally by a discussion/solution.

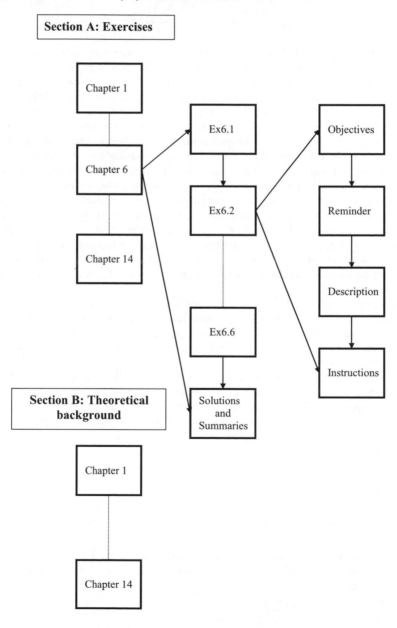

The exercises themselves are denoted as Exercise_i_q, where i is the topic number and q that of the exercise. For each exercise there is an objective, then a short reminder of the relevant aspects/equations. This is followed by a description, instructions and finally statements of the desired tasks.

Displayed Colors

Specific colors for displayed graphs are mentioned throughout the book. These references pertain to the colors displayed on the screen when running the exercises. The book itself is printed in black white, and these colors are obviously not visible in the printed version.

Suggestion of Use

For a basic minimal course, the material would be Chapters 1–5, 7 and 10. A more comprehensive program would consist of chapters 1–10 possibly adding chapter 11–12. Chapter 13 (and some exercises, for example ex3.3) while oriented to mechanical aspects demonstrate how *understanding the limitations* of specific processing techniques is essential in order to extract the relevant information.

Chapter 14 would be appropriate for students of general science (or physics).

CD and Website

The CD includes all the programs necessary to run the exercises, using the Matlab software. All necessary instructions to install and run appear on a readme file.
Note that an underscore will be used when running program, as in ex10_2

Inserting the CD will activate an autorun. However, Internet security may be active, shown by the appearance of the message 'To help protect your security, Internet Explorer has restricted this file from showing active content that could access your computer. Click here for options...' in the upper part of the IE, below the address bar. In such a case:
   a) Click the Information Bar.
   b) Click 'Allow Blocked Content'

The following website may be used in the future for possible updates and corrections:
www2.technion.ac.il/~braun

Acknowledgements

All exercises have been used in formal, one semester engineering courses, as well as 4 day workshops. I am deeply indebted to my students and workshop participants for their valuable feedback and suggestions, most of which are reflected in the final material. Many thanks to Dr. Michael Feldman, for his critique and help with Matlab programming, and to my daughter Orit for enlightening me on Microsoft's Word idiosyncrasies. And finally, thanks to my spouse Mira, who kept after me to finish this book, overcoming my usual tendency to leave unfinished as many tasks as possible.

About the Author

S. Braun is Professor at the Israel Institute of Technology, and has been active in the area of signal processing for 4 decades. He is the Editor in Chief of the Journal "Mechanical Systems and Signal Processing" (Elsevier), the " Encyclopedia of Vibrations" (Academic Press) and author of "Mechanical Signature Analysis" (Academic Press). He has authored more than 60 papers in peer-review journals and presented his work at conferences. He has conducted worldwide workshops (Europe, US, Australia, China), attended by academics and engineers and, during sabbaticals, applied his expertise to industrial problems at GE and Ford Motor Co.

Notation

Exercise Icon
This icon indicates every time an exercise is displayed.

Solution Exercise
This icon indicates every time a solution is displayed.

ADC	Analog to Digital Converter
AM	Amplitude Modulation
AR	Auto Regressive
ARMA	Auto Regressive Moving Average
AIC	Akaike Information Criterion
B, BW	Bandwidth
EU	Engineering Units
DAC	Digital to Analog Converter
DFT	Discrete Fourier Transform
DFTS	Discrete Fourier Transform Series
FFT	Fast Fourier Transform
FIR	Finite Impulse Response
FM	Frequency Modulation
FPE	Final Prediction Error
FRF	Frequency Response Function
FS	Fourier Series, Full Scale
FT, F[ ]	Fourier Transform
HT	Hilbert Transform
IFT, F^{-1}[ ]	Inverse Fourier Transform
IIR	Infinite Impulse Response
IR	Impulse Response
MA	Moving Average
MS	Mean Square
NB	Narrow Band
NR	Noise Reduction
PSD	Power Spectral Density
RMS	Root Mean Square
SDOF	Single Degree of Freedom
SISO	Single Input Single Output
S/N	Signal to Noise
S,s	Laplace domain, variable

STFT	Short Time Fourier Transform
TDA	Time Domain Average
Z,z	Z domain, variable
2DOF	2 Degrees of Freedom

Symbols/Notation

Er	Energy
e	Error
f, f_s	Frequency, Sampling frequency
H	Transfer Function, Frequency Response Function
N,NFFT	FFT Block-size
P	Power
R	Correlation
S	Power Spectral Density
T	Time span
W,w	Window functions
δ	Dirac function
Δf	Frequency resolution
Δt	sampling interval
γ^2	Coherence function
ζ ,z,zeta	Damping Ratio
ω	Frequency (Radial)
x	Lower case variable - in time
X	Upper case variable – in Transform Domain (Fourier, Z)
$\otimes$	Convolution operator
*	Complex conjugate (upper right star)
$\wedge$	Estimator (hat)
$||$	Absolute, modul

Part A

The Exercises

These exercises are divided into 14 chapters, each one addressing related topics. Each chapter includes a number of exercises. The relevant theoretical background is given in Part B of the book, divided into the same major 14 chapters.

All exercises appear in a similar format, stating the objectives and tasks to be performed. When deemed appropriate, a reminder of basic formulae or characteristics is presented, followed by descriptions of the graphical interface and controls. Each chapter then concludes with detailed solutions and summaries for all the exercises contained in it.

Most exercises have a similar formats, but standards are preserved only to a limited extent. As many facets were developed after interacting with, and testing by, students and workshop participants, many of the specific aspects of each exercise reflect their inclination and input. However, some controls are standard – specifically the availability of a help/instruction menu. This should be used whenever difficulties are encountered when attempting to perform a specific task.

Readers and potential users of this book may naturally have their own preference concerning the order of exercising interactively and/or studying the theoretical background. However, considering that the intent of this book is basically to enable interactive learning, some suggestions are hereby offered.

Based on the author's experience it is advisable to tackle the topics sequentially in the presented order, and apply a multiphase approach. For each chapter, get first briefly familiar with the theory, and then run all the exercises. For each exercise it is best to first get familiar with the controls and capabilities, and only then attempt to accomplish the described task. A first attempt to achieve them should then be made, followed by an in-depth study of the summary/solutions.

A second phase of investigating selected exercises is then recommended, checking whether the important aspects described are now mastered and understood. This can then be followed by an in-depth study of the theoretical background.

A final remark seems in order: while specific tasks are set down, many enable additional investigations and tests beyond the ones indicated. Hopefully readers will be inclined to undertake additional explorations, after having become familiar with the suggested exercises.

1

Introduction

Overview

One major aspect of signal processing approaches is the possibility of interpreting data in different domains like time and frequency. Depending on the purpose and signal properties, one domain may be preferable to the other. Often the rationale of using the frequency domain is not obvious to the newcomer.

These introductory exercises demonstrate cases where the interpretation of results in the frequency domain seems more intuitively related to the understanding of the physics involved. They assume only some very basic knowledge concerning Fourier decompositions. However for all readers, but especially those for whom the notion of 'Fourier' is completely new, it is recommended to run the exercises again after mastering spectral analysis (Chapter 7) and input/output identification (Chapter 11).

Exercise 1.1 deals with a system excited by a periodic input, whereas Exercise 1.2 deals with one excited by a transient.

The Exercises
Exercise 1.1

Objective

To interpret the response of a resonating system to a periodic excitation in the time and frequency domain.

Reminder

The response of a system excited by a periodic signal $x(t)$ is

$$x(t) = \sum_{k=0}^{\infty} X_k \cos(2\pi k f_0 t + \theta_k)$$

$$Y(\omega) = \left[\sum_{k=0}^{\infty} X_k \cos(2\pi k f_0 t + \theta_k) \right] H(\omega)$$

Discover Signal Processing: An Interactive Guide for Engineers S. Braun
© 2008 John Wiley & Sons, Ltd

Description

The system at hand is a single degree of freedom second-order system. Two parameters, the undamped natural frequency W_n and the damping ratio zeta, can be controlled via sliders.

The lower left and right plots in Figure E1.1 show the periodic excitation and responses respectively. The upper plots (left, excitation; right, response) show the frequency domain representation in the form of spectral plots. Each line represents a harmonic component, with amplitude and frequency as noted (frequency is in radian/sec). The time function shown in the lower plot, has been decomposed into the sum of harmonic functions, described by the spectral plots. (The additional information, that of the relative phase between each output and input component, is not shown here.)

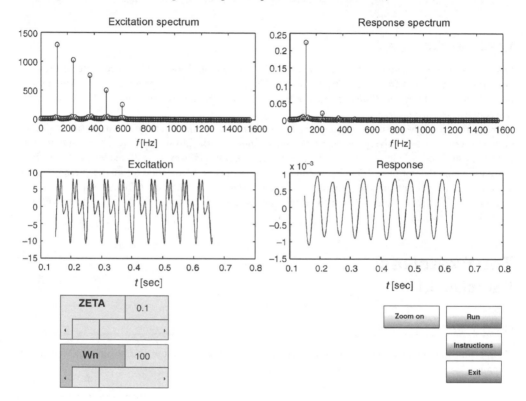

Figure E1.1

Instructions

Run the program for the following four cases:

Zeta = 0.04
W_n = 193

Zeta = 0.04
W_n = 480

Zeta = 0.01
W_n = 368

Zeta = 0.7
W_n = 368

It is recommended to run the program (with these four sets of parameters) twice. For each set of parameters inspect and note first the time domain plots only. Next note both the time and frequency domain plots.

Tasks

To summarize the possibility of interpreting intuitively all results by:

1. inspecting the time domain signals only;
2. to have the description in the frequency domain as well.

Exercise 1.2

Objective

To interpret the response of resonating systems to a transient excitation in the time and frequency domain.

Reminder

The response of a base excited mass–spring–dashpot system is

$$Y(\omega) = \frac{j\omega c + k}{-\omega^2 m + j\omega c + k} X(\omega) = H(\omega)X(\omega)$$

The shorter a transient, the wider the frequency range spanned by its spectrum.

Description

This exercise simulates the vertical displacement of a car, using a simplified model where rocking modes are ignored. The car travels over a bump having the geometrical shape of a half sine form. The speed of travel is relative to the parameter a, controllable via a slider; however, typing a number into the 'a' box and using return, allows more accurate control.

Two models of the suspension can be used, one of a single degree of freedom (SDOF) second-order system, the other of a 2DOF system with 2 degrees of freedom. The upper left plot in Figure E1.2 shows the excitation signal, the horizontal displacement of the wheel. The lower left one shows the response, the vertical displacement of the driver seat. The frequency descriptions are shown in the upper and lower right plots. The middle plot shows the frequency response of the system, which describes the gain as a function of frequency.

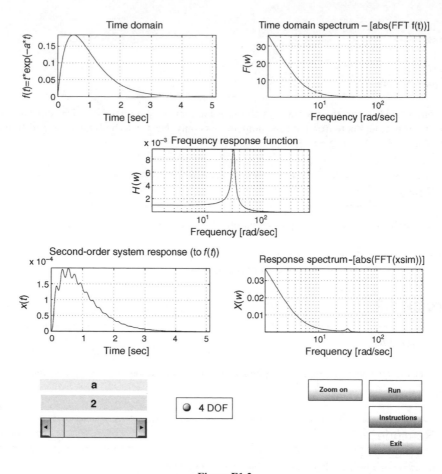

Figure E1.2

Instructions

Run the program for the values

$a = 2, 20, 100$

Repeat for the 2DOF mode, with $a = 50$.

Tasks

To summarize the possibility of interpreting intuitively all results by:

1. inspecting the time domain signals only;
2. having the description in the frequency domain as well.

Solutions and Summaries

Exercise 1.1

The fact that the excitations and responses are periodic can be noted by the spectral lines, showing frequencies which are multiples of a basic fundamental frequency (f = 19.5 Hz, W = 122.5 rad/sec). For zeta = 0.04, the following responses occur for W_n = 193 and 480 rad/sec (Figure E1.3)

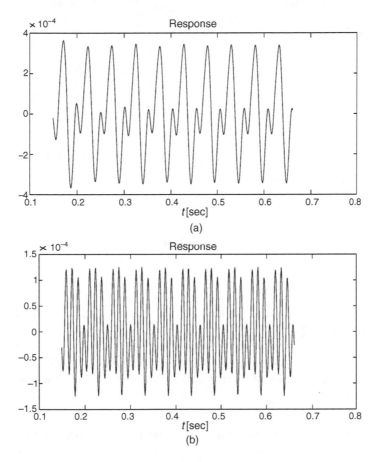

Figure E1.3

The results are very different, difficult to interpret from the signal plots only.

Changing W_n affects the peaks in the frequency domain plot. When W_n coincides with a multiple of 122.7 rad/sec, the relevant peak grows dramatically, and the response looks like a sine signal of the same frequency (Figure E1.4).

The third line of the excitation 368 rad/sec has thus been amplified. Whenever the resonant frequency coincides with one of the signal's components, the amplification occurring with small damping ratios results in this component dominating the output. For zeta = 0.7, almost no additional gain occurs at the resonance frequency; the excitation and response spectra (and hence the time signals) are then almost similar.

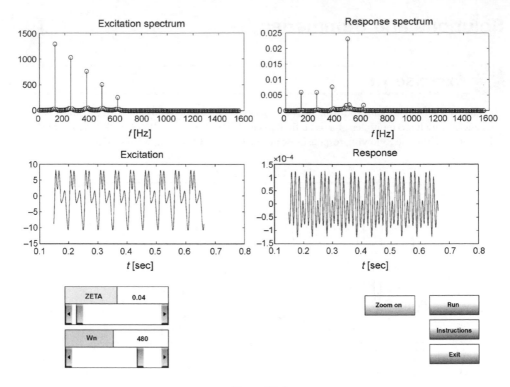

Figure E1.4

Exercise 1.2

For a low speed, $a = 2$, the response is almost similar to the excitation (Figure E1.5). For a higher speed, $a = 20$, the spectrum covers a larger frequency range. Furthermore, small oscillations appear in the output (Figure E1.6).

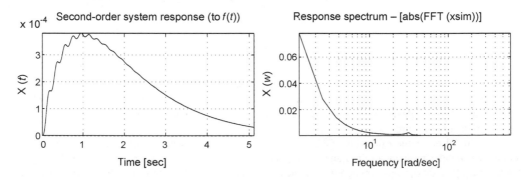

Figure E1.5

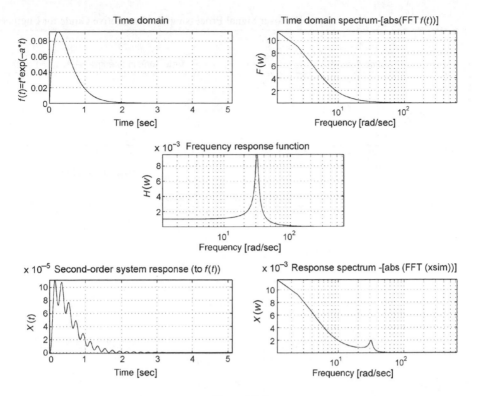

Figure E1.6

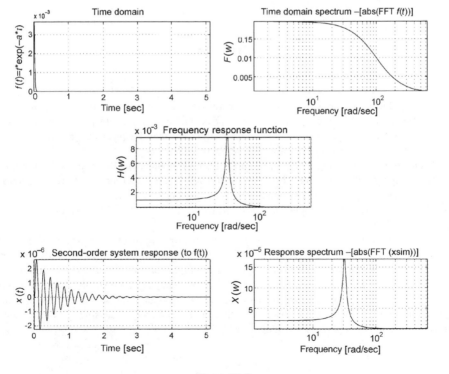

Figure E1.7

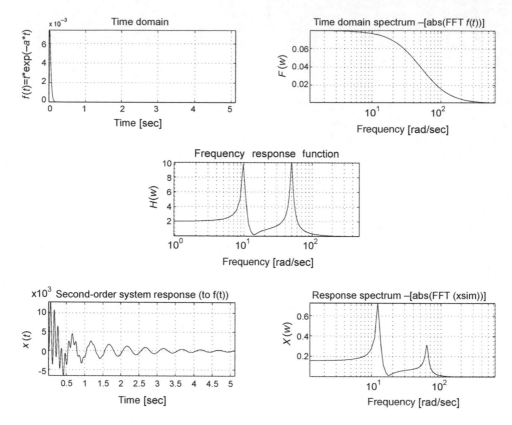

Figure E1.8

The system's frequency response function shows a strong peak at 30 Hz, indicating a resonance at this frequency. The excitation spectrum, covering a wider frequency band, now extends to the region beyond 30 Hz, the resonant frequency, and a small peak can be seen in the response spectrum, indicating energy at this frequency. Zooming in on these oscillations, their frequency is approximately 30 Hz. Hence they are indeed effects of the resonance being excited.

For a much higher speed (Figure E1.7), $a = 100$, it is now obvious that the response approximates to the impulse response of the system. This consists of a very short excitation, oscillating response with frequency concentrated around the resonant frequency of 30 Hz.

For the 2DOF system, the complementary information in the time and frequency domain is very helpful (Figure E1.8)

The excitation spectrum shows a lot of energy around the first resonant frequency of 30 Hz, less (but not negligible) energy around the second one of 60 Hz. The response spectrum shows a large peak at 30 Hz, a smaller one around 60 Hz. The response shows a faster oscillation at the beginning, which however decays very fast, and a second slower oscillation, decaying more slowly. The oscillating frequencies can be inspected by zooming in, showing values of 60 and 30 respectively. The fact that the higher frequency oscillations decay faster show that the loss of energy (causing the decay) somehow depends on the number of periods which occurred.

2

Signals

Overview

Applied signal processing methods must often be geared to the properties of the specific signals encountered. This chapter deals with signal classifications (Exercise 2.1), and introduces some of the basic parameters used to characterize them quantitatively (Exercise 2.2). The basic variability encountered with random signal is addressed by Exercise 2.3, and its effect when tracking changes by Exercise 2.4.

The Exercises
Exercise 2.1

Objective

To recognize signal classes.

Reminder

Signals can be classified as:

- Deterministic – periodic or transient
- Random – stationary or nonstationary

Also as:

- Power signal
- Energy signals

Discover Signal Processing: An Interactive Guide for Engineers   S. Braun
© 2008 John Wiley & Sons, Ltd

Description

Different types of signals can be inspected (Figure E2.1). The program responds immediately upon clicking the chosen category. Zooming may be necessary in order to investigate the signal patterns in more detail. For some cases, additional parameters or functions are depicted, i.e. mean, MS and RMS for periodic and random signals, positive and negative envelope for NB (narrow band) random signals.

The External Signal option handles signals generated by the file OUT_SIG.m

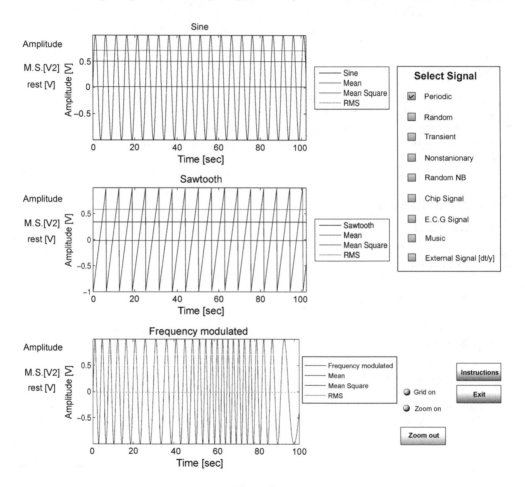

Figure E2.1

Instructions

This exercise involves mainly inspecting given signals. For some of the signals the parameters MS, RMS are shown. These should be ignored at this stage; they will be considered in the next exercise. For the narrow band (random) signal, the envelope is also shown. At this stage this should be understood as a concept only, an imaginary line 'enveloping' the signal's maxima. The notion of envelope will appear again later (Chapters 8 and 13).

Tasks

- For all cases shown, classify as power or energy signal.
- Check (and justify) the classification of all signals. For example, when choosing 'periodic', address also the lower plot.
- Explain the signal, mean, mean square and RMS values for the power signals.
- Describe (and compare) all the cases under the 'random' choice.
- Do the same for the transient and nonstationary cases.
- For the random NB case, describe the pattern obtained when zooming in a narrow horizontal strip (say ± 0.02 V) around the zero amplitude.

Exercise 2.2

Objective

To demonstrate the concept of energy and power signals.

Reminder

- Energy of a signal:

$$E = \int_0^{T_t} x^2(t)\mathrm{d}t$$

- Power of a signal:

$$P = \frac{1}{T}\int_0^{} x^2(t)\mathrm{d}t$$

for $T \to \infty$

$$\to \frac{1}{T_t}\int_0^{T_t} x^2(t)\mathrm{d}t$$

Description/Instructions

Energy (transient) or power (continuous) signals can be chosen (Figure E2.2). To choose a deterministic or random signal, the 'Continuous' option has to be chosen first.

The lower plots show the evolution of energy and power with time, while the total values are shown in the lower boxes.

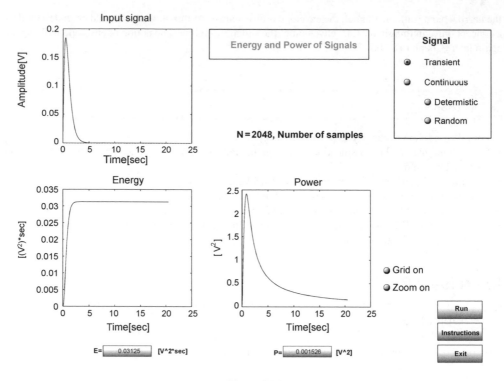

Figure E2.2

Tasks

Check the evolution of power and energy for transient signals, and note total value depicted.

    Repeat for continuous random signals. Compare the results when running this case more than once.

Exercise 2.3

Objective

To understand the concept of one 'realization' of a random signal. To investigate the variability of parameters or functions used to describe random signals.

Reminder

- A random signal of finite duration is considered as a single sample, or realization of a random process.
- Signal parameters like mean, variance or other statistical moments can only be estimated.
- Describing functions like correlation, spectra or distribution can only be estimated.
- Computed parameters or functions have a statistical variability from realization to realization.

Description

Parameters of the random signal are shown in the upper boxes, which give the values for the mean, mean square and RMS (Figure E2.3). The upper left plot shows the random signal investigated. The lower right plot is the probability density distribution. The middle and left lower plots are the autocorrelation and absolute Fourier transform squared (denoted as PSD – power spectral density), to be checked only after be coming familiar with the material of Chapters 3 and 7.

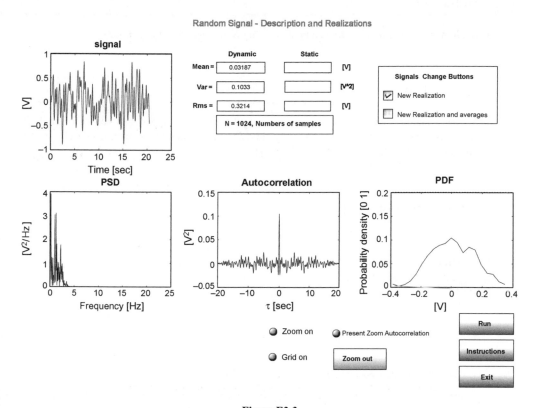

Figure E2.3

A new realization can be chosen for each run. The results shown (parameter and functions) will show some variability.

Prior average parameters and functions, based on 100 realizations, are available. These average results can be shown together with those corresponding to a new realization by choosing the option 'New realization and averages'.

Instructions/Tasks

Use the option of 'New realization', and run the exercise a multitude of times (at least 10). Investigate and describe qualitatively all the parameters (mean, mean square, RMS) as well as the statistical functions. For the autocorrelation function use also the preset zoom button.

Repeat using the 'New realization and averages' option.

Exercise 2.4

Objective

To investigate parameter variability, using examples of nonstationary random signals.

Reminder

The random error can often be reduced by increasing N, the number of data points.

Description

A nonstationary random signal can be chosen, checking the box in the upper right box of the analysis section. The signal can be divided into sections by choosing the number of data points in each section. The number of sections will then be set accordingly (Figure E2.4).

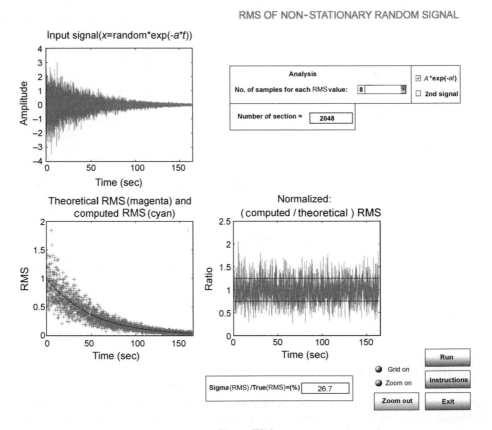

Figure E2.4

The RMS values are computed for each section, and shown on the lower left plot, superimposed on the theoretical instantaneous RMS value (which is obviously a function of time). An easier inspection of the variability of the RMS estimates is shown by the lower right plot. This depicts the normalized RMS value, the computed one divided by the theoretical value for this section. Theoretical error bounds, the red horizontal lines (by methods outside the scope of these exercises) are also shown. The computed mean of those errors is shown in the lower box.

Instructions

Choose one signal, run the program for various section durations. Repeat for the second signal.

Tasks

- Choose a reasonable sample, explaining the criteria for the choice.
- Determine (qualitatively, and possibly quantitatively) the effect of the sample size (signal length) on the random error (uncertainty) of the RMS estimate.
- Comment on the accuracy of the theoretical prediction of this error.

Solutions and Summaries

Exercise 2.1

Power signals are continuous in time; see periodic and random, narrowband random and some of the nonstationary signals. The signal's duration is chosen to span an arbitrary (practically) time length. Their strength can be described by parameters like mean square (note the nonzero mean and value of 0.5 for a unit amplitude sine wave) with [V^2] units, or its square root (0.707), the RMS, in [V] units.

The narrow band random signal has specific properties, observable by zooming in. As opposed to the wide band signal, it seems to have a 'constant' frequency, at least by defining a zero crossing period. Looking at the envelope, we may almost describe a slow random amplitude modulation of a harmonic signal (Figure E2.5)

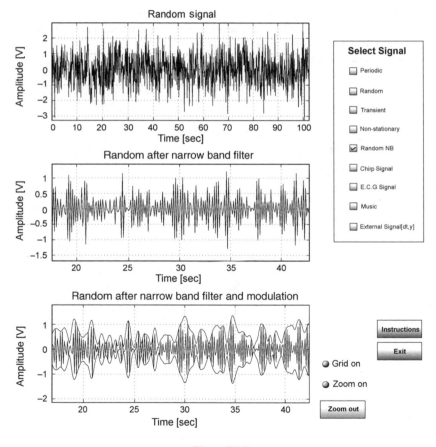

Figure E2.5

The transient signals have fixed durations and are energy signals. All those shown are deterministic. Nonstationary random signals can be of different types. The upper one seems to a random constant signal, but around a varying mean. The middle one has a slowly decreasing magnitude, but its basic shape is random. The lower has a changed pattern around 50 sec, still random but slower. Hence the formal types names, which appear as titles on the plots. Some nonstationarities, like the one shown, are easy to describe, more subtle ones may exist.

Exercise 2.2

For transient signals, the energy evolves in time until the complete duration has been spanned. Within the resolution available with a linear display, this is around 4–5 sec. The power tends towards zero, but would disappear only after an infinite time, obviously not computable for any real signal.

For deterministic (power) signals the energy grows as time evolves. The power tends to a constant value of 1.375 V^2 (Figure E2.6). For a sine wave, the power would be (amplitude^2)/2 [V^2], which is not the case shown. Hence this could not be a sine signal, as can be checked via zooming on the signal itself.

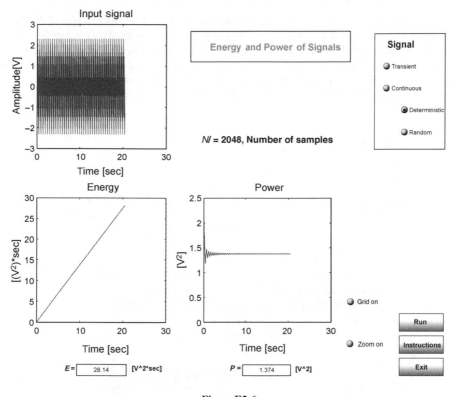

Figure E2.6

The power oscillates until it reaches a constant value, the fluctuations diminishing as the time evolves (i.e. the integration limit grows). A similar phenomenon occurs for random signals, except that the fluctuations no longer have the same typical pattern. Running the random case again (as per instructions, by pushing continuous and random buttons again) results in different signal patterns, and even power value. However, the final power value seems reasonably constant.

Exercise 2.3

Inspecting the signal itself (upper right plot), remarkable changes occur from one realization to another. Only some external knowledge, or possibly some tests, and certainly not the visual inspection alone could eventually convince us that all realizations are from the same stationary process.

Variability exists between statistical parameters (mean, variance, RMS) of different realizations. For example, taking the RMS, we note fluctuations spanning values from 0.345 to 0.288. The same exists for describing functions. We note, however, that the variability in shape and magnitudes is significantly larger for the PSD than for the correlation or distribution function.

Comparing with the average values (the option 'New realization and averages') we can compute normalized variabilities. These are far from negligible. Taking the mean, the full range of variations, normalized by the average, would be $(0.345-0.288)/0.335$, approximately 17%. An error figure for functions would necessitate a measure, but it is especially the PSD function which is problematic.

Exercise 2.4

The error between the theoretical and computed RMS can be seen. This is an example of the statistical error encountered in Exercise 2.3, showing variability in parameters extracted for random signals. In this exercise, the available data duration is fixed, hence a compromise may be necessary in the analysis.

The choice of signal length (in this case the number of data used for each RMS computation) depends on two conflicting criteria. The first is to track the RMS evolution, the second is to maximize the accuracy of the result (Figure E2.7). Running signal lengths from 8 to 2048, we note

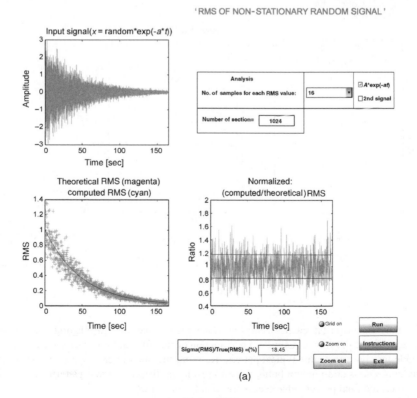

(a)

Figure E2.7

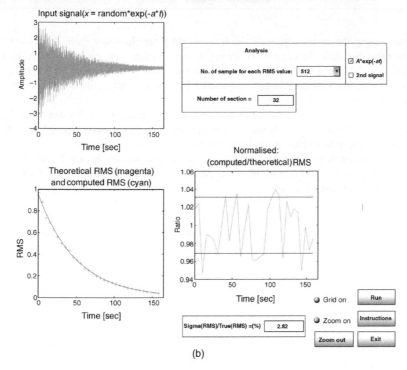

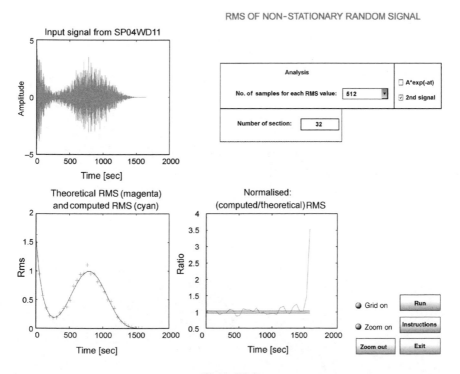

RMS OF NON-STATIONARY RANDDOM SIGNAL

(b)

Figure E2.7 (*continued*)

RMS OF NON-STATIONARY RANDOM SIGNAL

Figure E2.8

errors bands (using the zoom option on the lower right plot) changing from approximately 25 to 2%. On the other hand, at the extreme, only eight sections are then available for RMS computations, and the evolution of the RMS cannot be tracked with good temporal resolution. A choice of 256 section lengths, resulting in 64 computed RMS values and an error of 4 %, seems like a reasonable compromise.

The error seems inversely proportional to the signal length (data points). The theoretical prediction can serve as a good indicator, but will naturally not predict the error exactly (theory can never predict exact outcomes of computed values from random data).

The necessary compromise is very obvious for the second signal (Figure E2.8). Any choice of number of sections less than 16 (out of the available options) would not enable us to follow clearly the temporal evolution of the RMS value.

3

Fourier Methods

Overview

The majority of frequency domain approaches are based on Fourier methods, and understanding these is a must for most of the following chapters in this book. Exercises 3.1 and 3.2 introduce Fourier series, and their application to an experimental case is shown by Exercise 3.3. The case of modulated signals (Exercise 3.4) is important for those wishing to understand rotating machinery signals (Exercise 9.2 and Chapter 13).

Exercises 3.5–3.8 deal with Fourier transforms, and especially the uncertainty relations between time and frequency representations (Exercise 3.6). Digital processing, based on the fast Fourier transform (FFT), are then dealt with by Exercises 3.9 and 3.10. The basic computational frequency resolution is introduced by Exercise 3.9, and the introduction of engineering units (EU) in 3.10. Exercises 3.11 and 3.12 attempt to introduce the leakage phenomenon. Understanding of these topics is essential in applying spectral analysis (Chapter 7).

The Exercises
Exercise 3.1

Objective

To discover the Fourier series.

Reminder

For a periodic function

$$x(t) = x(t + T_p) = \frac{C_0}{2} + \sum_{k=1}^{\infty} C_k \cos\left(\frac{2\pi kt}{T_p} + \phi_k\right)$$

with T_p the period.

Discover Signal Processing: An Interactive Guide for Engineers   S. Braun
© 2008 John Wiley & Sons, Ltd

Description

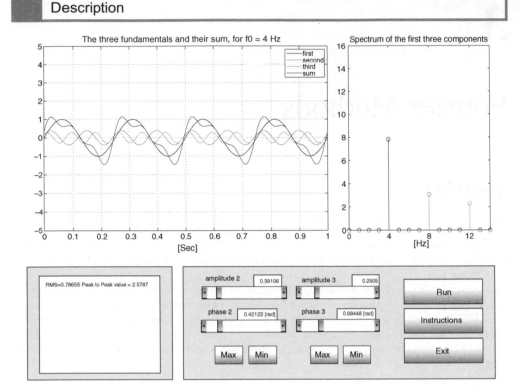

Figure E3.1

In this exercise we always start with a sine signal of frequency 4 Hz and amplitude of 1. Two additional harmonic signals, denoted by 2 and 3, can be added to this sine signal. Signals 2 and 3 have frequencies of 8 and 12 Hz respectively. Their amplitudes and phases (relative to the sine) can be controlled independently by means of the relevant sliders.

Shown in Figure E3.1 (left plot) are the basic sine signals, signals 2 and 3, and the sum of the three signals. The right plot shows the (amplitude) spectrum of the total signal. The RMS and peak to peak values of the total sum are shown on the lower left box.

Instructions

- Set amplitudes and phases of signals 2 and 3 to zero, inspect the signals, spectrum, RMS and peak-to-peak value.
- Next, check the effect of varying only amplitude 2 on the total signal. Repeat by varying also phase 2.
- Next, setting amplitude 2 and phase 2 to zero, check the effect of varying only amplitude 3 on the total signal. Repeat by varying also phase 3.
- Next, with phases 2 and 3 zero, vary both amplitudes 2 and 3.
- Next, choose arbitrary combinations of all amplitudes and phases. First modify only amplitudes, then only phases, and finally all four parameters.
- Finally, try to change only phases. Try various possibilities of chosen but fixed amplitudes), such that the positive, negative or peak-to-peak values are extreme (maximum or minimum).

Tasks

It is required to find a fixed characteristic of the total signal which exists for *any combination*.

Formulate a theorem showing a general property of the resultant signal for *all* the cases tested. Discuss the effect of varying the phase of the additional components.

Exercise 3.2

Objective

For the Fourier series decomposition of a periodic signals, to investigate the convergence rate. Also to see the Gibbs effect at the discontinuities.

Reminder

A symmetrical square signal (amplitude 1, zero mean, discontinuities at 0, $T_p/2$, T_p, etc.) has a Fourier series decomposition of

$$x(t) = \frac{4}{\pi}\left[\sin(\omega_o t) + \frac{1}{3}\sin(3\omega_o t) + \frac{1}{5}\sin(5\omega_o t)...\right], \ \omega_o = \frac{2\pi}{T_p}$$

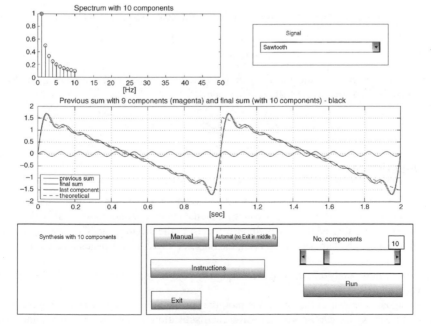

Figure E3.2

Description

Two types of periodic signals (square and saw tooth) can be analysed. The Fourier series decomposition can be undertaken manually: when the manual button is pressed, the next harmonic is added to the composite signal. The added harmonic, the previous and current sum as well as the signal to be decomposed are shown in the lower plot (Figure E3.2). The upper plot shows the amplitude spectrum of the current sum. The current number of harmonics is shown in the box next to the number of component slider, and also in the lower left box. The harmonics can be added automatically, resulting in a dynamic display.

Instruction/Tasks

- Choose a square signal, set number of components to zero (by overwriting zero in the slider box). Manually add harmonics. Inspect and evaluate results.
- Reset number of components to zero and operate automatically.
- Repeat for the sawtooth signal. Summarize the result.

Exercise 3.3

Objective

To analyze a real experimental case of a periodic signal via a Fourier series representation.

Reminder

The Fourier series components are computed at frequencies which are integer multiples of the basic frequency

$$1/T_{total} = f_{fundamental}$$

For a harmonic signal of frequency f_p, spanning exactly p periods in the analyzing window, the location on the frequency scale will be at the index

$$k = p$$

Description

The following information concerns the acquired data (choose the belt option in this exercise).

A rotor supported by two bearings is driven by a toothed belt. The data shown and analyzed consists of the reaction force acting on the bearing house nearest to the driving belt. The rotating speed is 240 rpm. The number of belt teeth per rotation is in the range of 11–12 (this depends on the rotor and radius and belt length).

Two signals can be analyzed, the experimental data from the rotating device, and a simulated sawtooth signal.

The raw signal is shown in the upper left plot, the spectrum in the upper right (Figure E3.3). The number of Fourier series can be set manually, each activation of the manual button increasing the number of harmonics by one. The spectrum of the total signal (blue) is shown in the upper right plot,

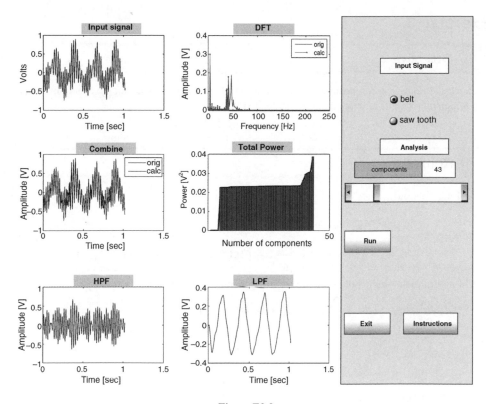

Figure E3.3

and superimposed on it the identified harmonics (red). The same information appears as time domain signals in the middle left plot. This shows the total signal (black) and superimposed on it the Fourier series synthesis, i.e. the signal composed of the sum of the components identified. The power of each component (black bar plot) as well as the total accumulated power (blue line) appear in the middle right plot. The program can run in automatic mode, with harmonic order increasing from 1 to maximum, and can be paused (continued) by the appropriate buttons.

The lower plots result from filtering, an operation addressed in Chapter 5. The slow (low frequency) and fast (high frequency) components are shown by the right and left plots respectively.

Instruction/Tasks

Inspect the raw signal; it should be possible to give a physical interpretation (roughly) for the time domain pattern, and specifically to the slow and fast oscillations shown. Step manually through the Fourier decomposition, and try to relate the time pattern of the reconstructed time signal to the specific harmonics added in the decomposition/synthesis. Inspect also the power figure information. It is necessary to determine which harmonics describe specific physical phenomena, which describe unidentified fluctuations, how many harmonics are needed in order to describe the basic character of the signal, and a quantitative value may be associated with these identified patterns.

Exercise 3.4

Objective

To check the Fourier decomposition of harmonic signals modulated by harmonic functions. This will help understand results of gear signals (Chapter 13).

Reminder

AM modulation:

$$x_{am}(t) = A\left[1 + k_{am}\cos(2\pi f_m t)\right]\cos(2\pi f_c t)$$

$$x_{am}(t) = A\cos(2\pi f_c t) + \frac{Ak_{am}}{2}\left\{\cos\left[2\pi(f_c + f_m)t\right] + \cos\left[2\pi(f_c - f_m)t\right]\right\}$$

FM modulation – higher order (symmetric) sidebands are generated:

$$x_{pm} = A\sum_{k=-\infty}^{\infty} J(k_{pm})\cos\left[2\pi(f_c + kf_m)t\right]$$

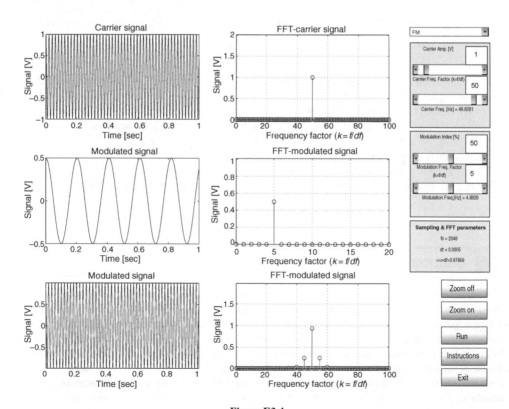

Figure E3.4

Description

The user can choose (via a pop-up) cases of amplitude or frequency modulation (AM or FM), where the carrier frequency and amplitude, the modulating frequency as well as modulation depth can be controlled. An additional case of combined AM and FM is available, with all parameters constant. When choosing this case, a return to AM or FM only is done via a push button. For all cases, the carrier and modulation signals are harmonic.

Shown in Figure E3.4 are the carrier, modulation signal and modulated signals, with the time domain in the right plots, and spectra in the left ones.

Instructions/Tasks

- Choose AM, change the modulation frequency and amplitude, and determine the effect on the spectra.
- Repeat for the case of FM.
- Choose the case AM% FM, and note the parameters used. Choose AM only with these parameters (those relevant to the AM case), next FM only with these parameters (this relevant to the FM case). Compare with the case of AM and FM.

Exercise 3.5

Objective

To examine the possibility of synthesizing a transient (aperiodic) signal by harmonic components.

Reminder

$$x(t) = \int_{-\infty}^{\infty} X(f)\exp(j2\pi f)df$$

$$X(f) = TX_k = \int_T x(t)\exp(-j2\pi ft)dt$$

Aperiodic functions can be synthesized and decomposed by harmonic functions, spanning a continuous range of frequencies.

Description

Two types of transient signals to be synthesized can be chosen, as well as the number of synthesizing components.

A dynamic display is updated as each additional component is added, until the maximum number is used. The upper left plot in Figure E3.5 shows the desired final signal, and the sum of components at each step. The other plots shows all generated components used: all components appear in a vertical display (upper right), appended to each other (lower right) and (dynamically), the generated components themselves.

Remark: No continuous range of frequencies can be simulated by a digital computer. This exercise mainly attempts to hint at the possibility of generating aperiodic functions via finite summations of harmonic functions. The actual computation, by necessity, performs discrete summations.

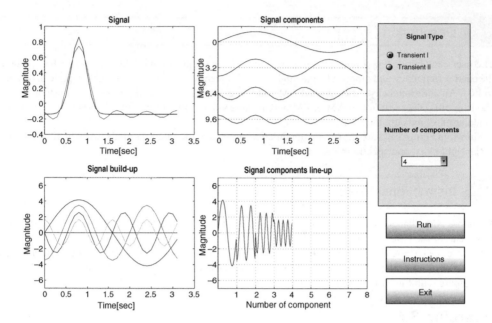

Figure E3.5

Instructions/Tasks

- Choose transient 1, observe the effect of synthesizing it with an increasing number of components.
- Repeat for transient 2.

Exercise 3.6

Objective

To investigate the Fourier transform scaling theorem.

Reminder

For $x(t) \leftrightarrow X(f)$ and a scalar parameter a

$$x(at) \leftrightarrow \frac{1}{|a|} X\left(\frac{f}{a}\right)$$

Description

Running the exercise shows a first window with the signal (left plot) and the (absolute value) Fourier transform (right) (Figure E3.6). A slider enables us to compress or expand the signal, the resulting energy being shown.

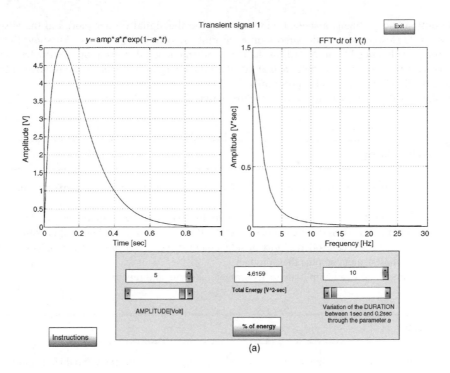

(a)

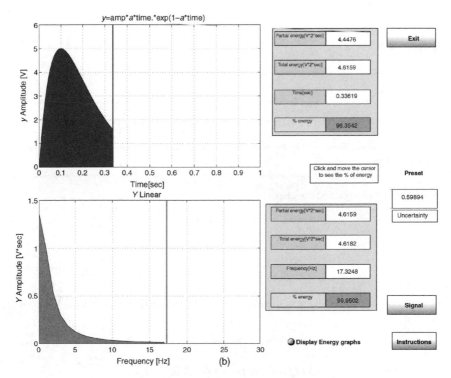

(b)

Figure E3.6

The button energy opens a second window, showing the signal (upper plot) and the absolute value of $X(f)$ (lower plot), with cursors on each plot. Scrolling the cursors will show the area under the plots bounded by it. Noted for the time and frequency domain plots are the location of the cursors, the energies of the bounded areas, the total energies and finally the relative energies in percent. (Read instructions for program)

The signal button enables us to return to the first window.

Instructions

In the first window modify amplitudes and signal durations. Find the qualitative effect of the signal duration in both domains.

Next set the amplitude to 3 and the parameter a to 10, switch to the second window, compare the energy in both domains. Then move cursors such that 50% of the energies is bounded in both domains (note that this can only be controlled approximately; do not strive for absolute accuracy). Note the time and frequency chosen, and compute their product.

Return to first window, set amplitude to 3, parameter a to 50, and repeat.

Task

Investigate the effect of the signal duration on its FT. Investigate the time–bandwidth product for various signal durations.

Exercise 3.7

Objective

Understand the convergence from Fourier transform to Fourier series for periodic signals.

Reminder

- For a finite duration periodic signal,

$$x_T = x(t)\, w(t)$$

$$x(t) = x(t + T_p)$$

$$w(t) = u(t)$$

with u the step function.
- For a periodic harmonic function, of unit amplitude, spanning time T, then from Equation (3.6) in Part B,

$$T \rightarrow \infty$$

$$|X(f)| = 0.5\, \delta(f - f_0) + 0.5\, \delta(f + f_0)$$

Three types of periodic signals can be chosen. A truncated signal can then be chosen, spanning a controllable number of periods. Running the program results in a dynamic display, as one additional period is spanned at each phase, until the preset number of periods has occurred. At each phase, the time signal (upper plot) and the magnitude (modul) of the FT (lower plot) are shown (Figure E3.7).

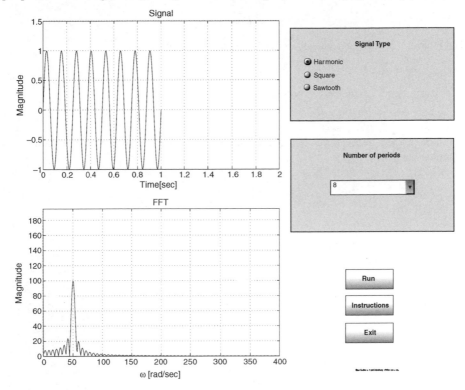

Figure E3.7

Instructions/Tasks

Check the effect of increasing the number of periods spanned. Repeat for each of the signal types available.

Exercise 3.8

Objective

For a harmonic signal, see the relation between the number of periods scanned, and, computing an FFT, the index of the spectral line.

Observe the relation of the spectrum of a single period of a periodic signal, to that of its truly periodic extension.

Reminder

For FFT-based spectra, the frequency spacing is

$$\Delta f = \frac{1}{N\Delta t}$$

From an FFT computation

$$X(k) \leftrightarrow x(n)$$

The Fourier transform, at frequencies $k\Delta f$, is approximated by

$$X(f) \approx \Delta t \, X(k\Delta f)$$

The Fourier series coefficients are computed as

$$X_k = \frac{1}{N} X(k)$$

Given the FT of a transient signal, the spectrum of the periodic signal (with the given transient spanning one period $1/f_0$) is

$$X(k) = \frac{1}{N} X(f), \quad f = kf_0$$

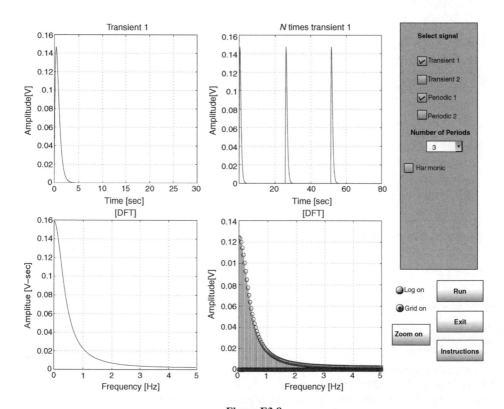

Figure E3.8

Description

Two possible options are available for the choice of signals: (1) a transient together with a periodic repetition of it, or (2) a harmonic signal. For the first case, a choice of two different transients is available, and the number of repeated transients can be controlled. The number of periods spanned by it is controllable for the harmonic signal. Shown in Figure E3.8 is the frequency resolution of the FFT analysis for this case. The signals are shown in the upper plots, and the spectra in the lower plots. The lower plots use a normalization factor, such that the spectra are independent of the number of periods used. For case 2 only one time and frequency domain plot are needed.

Instructions/Task

Choose a harmonic signal. Choosing $K = 4$ determine the number of periods spanned by the signal, as well as the index of spectral line (zoom may be necessary). Repeat for K from 5 to 9.

Choose signal transient 2. Extend the number of periods from 2 and on, compare the spectra at all computed frequencies for the aperiodic and periodic signal. Repeat for transient 2.

It is required to determine the relation of the spectra of a transient to is extended periodic signal:

1. qualitatively, concerning the resultant spectral shapes;
2. quantitatively, concerning the frequencies shown.

Exercise 3.9

Objective

To see the relations between Δt, NFFT and Δf, for DTF computations via the FFT.

Reminder

$$\Delta f = \frac{1}{N\Delta t} = \frac{f_s}{N}$$

$N\Delta t$ is the signal duration.

Description

Shown in Figure E3.9 is an exponentially decaying signal (left plots) and its (absolute) DFT (right plots). Theoretical values are given by a continuous green line, the (blue) stem with a red circle on top correspond to discrete times or frequencies (computed via the FFT) samples. The upper plots show results using a fixed sampling interval of 5 [msec], the lower ones a fixed signal duration of 6.4 [sec]. N, the number of samples used for the DFT calculation, is controllable.

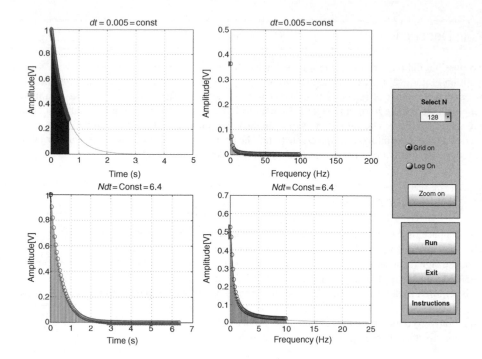

Figure E3.9

Instructions/Tasks

Determine the effect of changing N when:

1. The sampling interval is constant;
2. The signal duration is constant.

Use the zoom option in order to examine closely spaced results. Use the log option when the behavior for small values of the DFT is to be checked.

Exercise 3.10

Objective

To understand engineering units for one- and two-sided spectra, according to the signal class.

Reminder

The DFT of an N point sequence $x(n)$ is

$$X(k) = \sum_{n=0}^{N-1} x(n)\exp\left(-j\frac{2\pi}{N}nk\right)$$

with

$$X(-k) = X(N - k)$$

Negative frequencies are associated with $k > N/2$ for a two-sided transform $X(k)$. A translation of the section comprising the last $N/2 \ldots N$ transform values to the negative frequency region, to give the more classical symmetrical representation, with $N/2$ the Nyquist frequency. A one-sided representation, from 0 to $N/2$, may be used, with an appropriate change of magnitude – a factor of 2, except at end points 0 and $N/2$.

$X(k)$ being complex, symmetries occur for real signals $x(n)$. Representations of magnitude, real, imaginary and complex displays are possible.

Engineering units will differ according to signal types:

Aperiodic: DFT   [V-sec]
Periodic: FS   [V]
Random: PSD   [V²/Hz]

Description

The DFT pairs $x(n)$ and $X(k)$ are shown in the upper left and right plots in Figure E3.10. The FFT, two- and one-sided spectral representations are shown in the left, middle and right lower plots.

Three signal types can be chosen. The relevant EUs appear in the lower plots.

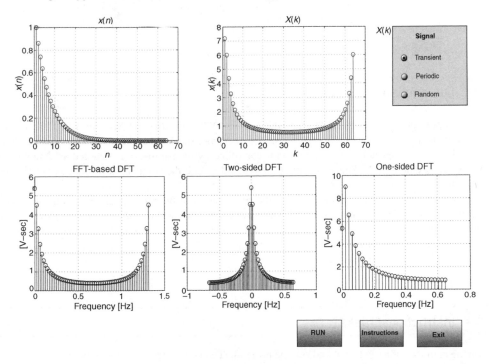

Figure E3.10

Instructions/Tasks

For all signal types, note the EU of the spectra. Next check the quantitative aspects of these, as related to the DFT pair $x(n)$ and $X(k)$. Finally observe the differences of the one- and two-sided representations, checking those of end points as well.

Exercise 3.11

Objective

To demonstrate the periodicities inherent in the DFT. This is a preliminary exercise, needed in order to understand Exercise 3.12

Reminder

The DFT pair is

$$X(k) = \sum_{n=0}^{N-1} x(n) \exp\left(-j\frac{2\pi}{N}nk\right)$$

$$x(i) = \frac{1}{N}\sum_{0}^{N-1} X(k) \exp\left(j\frac{2\pi}{N}nk\right)$$

Description

The two upper plots in Figure E3.11 show the N point DFT pair $x(n)$ and $X(k)$ for a specific signal. The lower right plot shows the computation of the $X(k)$ from the N point sequence $x(n)$, via the DFT, when k is extended beyond N. The extension is controlled by the parameter p, denoted as 'Number of periods', and equal to $p*N$. Thus choosing p = 4 would extend the summation in the computation of $X(k)$ to $4N$.

In a similar way, the lower left plot shows the computation of the $x(n)$ from the N point sequence $x(n)$, via the inverse DFT, when it is extended beyond N. The extension is controlled by the parameter p, denoted as 'Number of periods', and equal to $p*N$. Thus choosing p = 4 would extend the summation in the computation of $x(n)$ to $4N$.

The actual N can also be chosen.

Instructions/Tasks

- Change number of periods to 4 and run the program. Then check the effect of changing N.
- Repeat for variable number of periods.
- Finally check the effect of using a log scale.

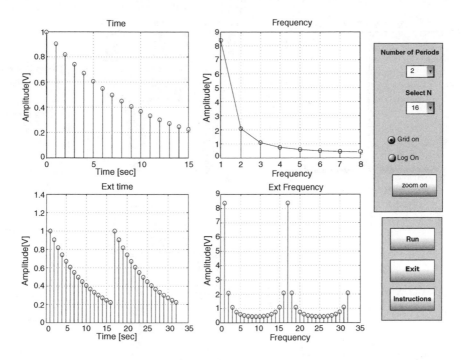

Figure E3.11

Exercise 3.12

Objective

To demonstrate leakage as due to a discontinuity, and its possible control by windows.

Reminder

Discontinuities in the time domain result in high frequency components. The solution: multiplying the time data by a time window which gradually attenuates the beginning and end of the data in the analysis window.

$$x(i) \rightarrow w(i)\, x(i)$$

Description

A harmonic signal of N samples is generated, with a chosen number of half periods (3 in Figure E3.12). The FFT (upper left) is computed. Repeating the last exercise (3.11), the indices n and k of the time and frequency DFT sequences are increased, N being multiplied by a factor chosen under the pop-up 'number of sections'. In the figure this factor equals 2, and with number of half periods in each section chosen as 3, the DFT covers $2N$ samples in both domains. In the above case, with a noneven number of half periods per section chosen, a discontinuity is generated in the time sequence.

The two lower plots show the effect of multiplying the first N points of the time sequence by a (Hanning) window, decreasing the effect of the discontinuities.

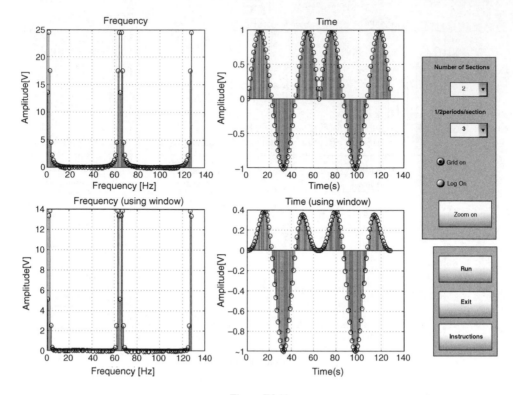

Figure E3.12

Instructions

Run the exercise with varying number of sections and number of half periods per section. Check the effect of applying a window, using linear as well as log scales.

Tasks

Summarize the effect of windowing the time signal when computing the DFT. Discuss the specific effect when analyzing a harmonic signal.

Solutions and Summaries

Exercise 3.1

For the original sine (Figure E3.13), a single line is shown in the spectrum. Adding the signal 2, the existence of a double frequency can be noted (for amplitude 2 big enough) by visual inspection. The phase has, however, a significant effect on the shape of the sum. Similar effects are observed when adding signal 3 (only). The amplitude and frequency can be very conveniently observed in the spectral representation. A more complex sum occurs when all four parameters are varied. However, *the basic period of the total signal is always constant!*

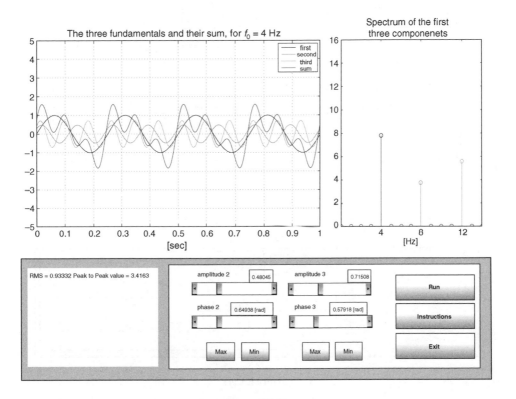

Figure E3.13

Thus we discovered the Fourier series. The synthesis consisted of adding to a fundamental component (the 4 Hz signal) two harmonics (of 8 and 12 Hz). The Fourier series description can thus decompose the periodic signal (the sum) into a fundamental and harmonics (whose frequencies are integer multiples of that of the fundamental).

The phase (obviously) is not reflected by the amplitude spectrum. The original signal cannot be recovered by the amplitude spectrum only. The phase does not affect the RMS values, but can have a significant effect on the peak-to-peak values of the total sum. Given the same RMS, the phase could be chosen such that the signal extreme values are maximized or minimized. This can be checked with the Max Min buttons.

Exercise 3.2

The convergence is relatively fast. The final shape of the periodic signals is recognized with approximately four to five harmonics (Figure E3.14). This is also obvious from the spectrum, as little power is added by higher-order harmonics. Checking the formula of decomposition for a square signal (see Reminder), we could note the fifth harmonic (third component) is down to 1/5 of the amplitude (and hence to 1/25 of the total power), the diminishing return of adding further harmonics being obvious. The convergence is problematic near discontinuities. Oscillations and overshoot (the Gibbs effect) are clearly seen for the square wave. No improvement occurs if additional harmonics are added. The convergence exists, however, for the mean of the oscillations.

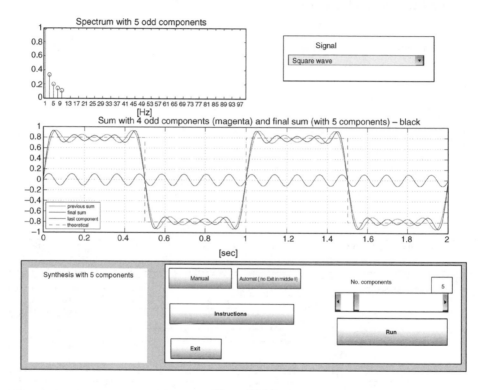

Figure E3.14

Exercise 3.3

Stepping through the number of harmonics, the fundamental component is recognized as the low-frequency pattern.

As seen in Figure E3.15, the time window spans 1 sec, with four periods, corresponding to a period of 0.25 sec, a frequency of 4 Hz, equal to the rotational frequency of 240 rpm. The fourth to sixth index of the FS corresponds to the approximate number of periods. The total power seems to stay constant until we get to the region of the thirty-second to fiftieth harmonic. The total power increases significantly, and the spectral picture shows that this increase is due to peaks in the region of 50 Hz. The time function shows that the high frequency fluctuations are now contained in the synthesized signal. Hence we recognize the low and high frequency components in the time, frequency and total power figures.

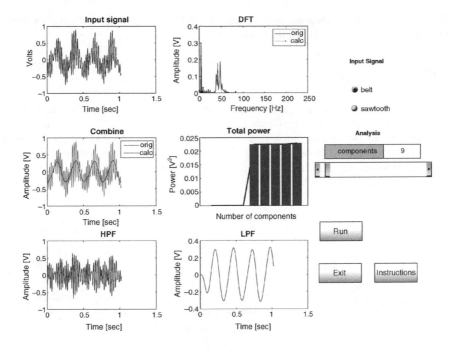

Figure E3.15

The belt tooth frequency being in the range of 44–48 Hz (11–12 teeth per rotation), the reaction force generated by these teeth could be modulated by the rotational speed of 4 Hz, generating power in the range of 40–52 Hz, the one shown by the high frequency region in the spectrum (Figure E3.16).

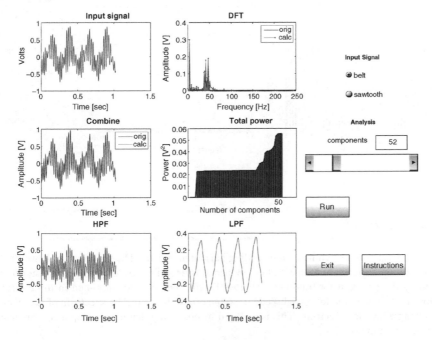

Figure E3.16

As the spectral regions are quite distinct, we could have done the separation of the components via filtering (see Chapter 5), as noted from the lower plots.

Exercise 3.4

For AM, we get one pair of symmetrical sidebands. The carrier component in the modulated signal is independent of modulation parameters (Figure E3.17).

For FM (Figure E3.18) we get symmetrical sidebands, but higher order sidebands can occur for higher modulation amplitudes. Both carrier and sideband components of the modulated signal depend in a complex manner on carrier and modulation parameters. For combined AM&FM, asymmetry of the sidebands can occur, as shown by this specific case.

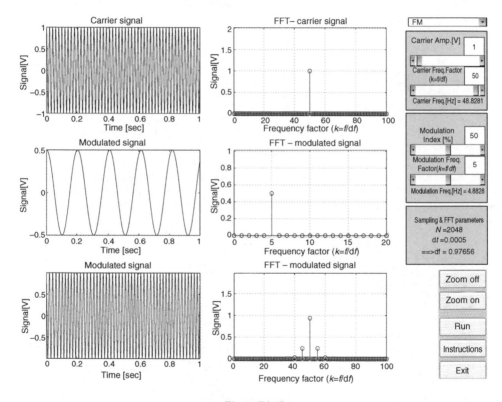

Figure E3.17

Exercise 3.5

Continuous transients can be synthesized by adding harmonic signals. The spectrum is, however, continuous, a continuous spectrum of spectral components as described by the Fourier transform. This is simulated by the program for transient 1 (Figure E3.19). The components seen in three different displays (lower left, upper and lower right) add up to the original signal. The spectrum of transient 2 is different, having a broader spectrum. This is deduced by observing that more components are needed for convergence to transient 2.

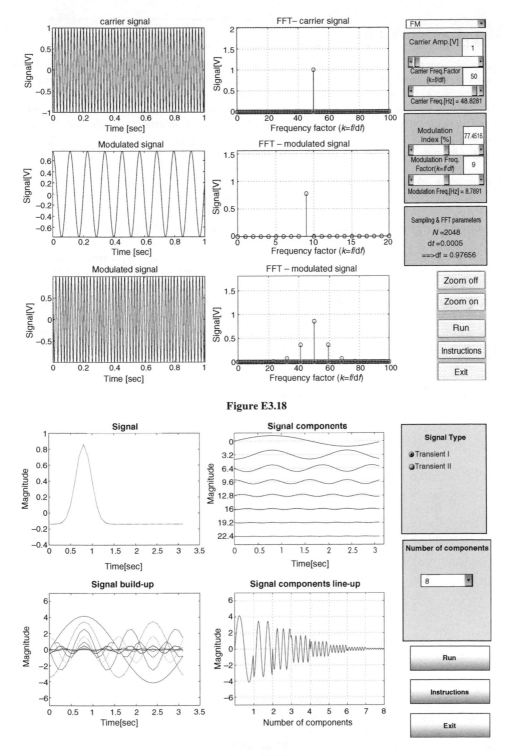

Figure E3.18

Figure E3.19

Exercise 3.6

Compressing (or expanding) in the time domain results in an expansion (or compression) in the frequency domain. Thus the shorter a pulse duration, the wider the spectral region covered (Figure E3.20). The total energy is, however, always equal in both domains.

In spite of the difficulty of controlling the cursors position exactly in order to bound a specifically desired relative energy, the results are consistent with the uncertainty product law for the FT (Figure E3.21).

Using the preset case (energy values chosen), we get:

- Case 1: $a = 10$, time cursor 0.346, frequency cursor 3.56 Hz, the product (noted in box) 1.235;
- Case 2: $a = 50$, time cursor 0.069, frequency cursor 16 Hz, the product 1.113.

Another test gave the following approximate (as cursor choice may be difficult) results:

- Case 1: $a = 10$: time cursor set at 0.18 (% energy 70), frequency cursor 1.6 (% energy 80).
- Case 2: $a = 50$: time cursor set at 0.038 (% energy 72), frequency cursor 6.9 (% energy 73).

The respective products of the durations $t_d * t_f$ are 0.288 and 0.26 respectively. We note, however, the low sensitivity of the energy function to the cursor location, and hence the high uncertainty in the computed products. An exactly invariant product can thus not be obtained, due to limitations of the program. The results obtained are still consistent with the invariance of the time–bandwidth product.

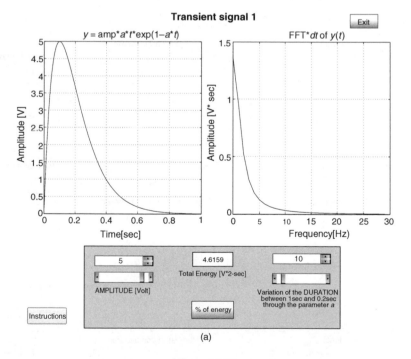

(a)

Figure E3.20

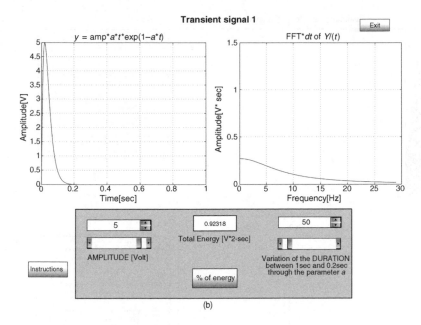

(b)

Figure E3.20 (*continued*)

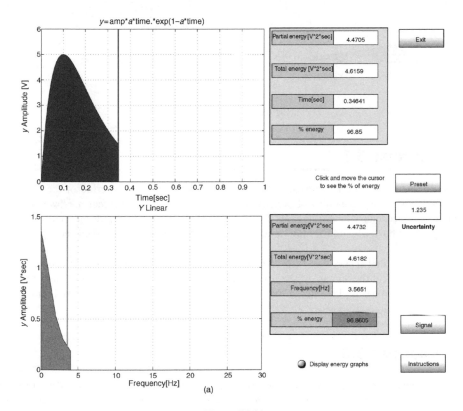

(a)

Figure E3.21

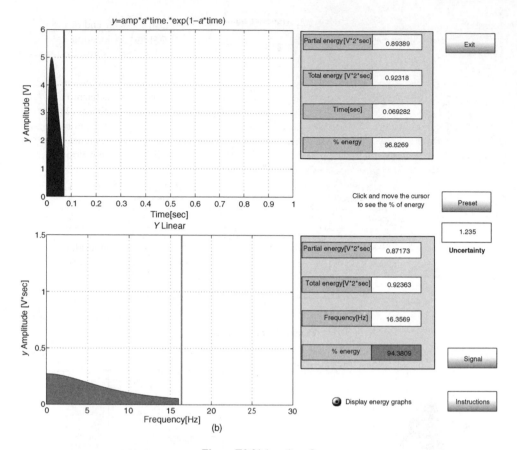

Figure E3.21 (*continued*)

Exercise 3.7

For a harmonic signal, the FT consists of a main lobe centred around 1/period of the signal, with existing side lobes. As the number of periods increases, the main lobe becomes narrower and larger. The number of secondary lobes increases, but the general spectra tend towards an impulse. This will obviously occur only when an infinite number of periods is spanned, i.e. when the signal is truly periodic.

For the square and triangular case, the same phenomenon occurs, repeated around the harmonic frequencies of the signal. At the limit, the FT would be a set of impulses. The (time) truncated square signal has higher energy around harmonic frequencies than that based on triangular signals.

Exercise 3.8

For the harmonic signal: the higher the number of periods spanned, the larger the index of the spectral line. Conversely, the relative location of a spectral line is an indication of the number of periods spanned by the signal – a large number of periods for locations towards the higher end of the spectral representation. For the case where an integer number of p periods is spanned, the location index of the single spectral line is (theoretically) $k = p$. As Matlab indexes start with 1 (and not zero), the true index is that appearing on the graph -1.

The line spectra of a periodic signal have an envelope tracking the FT of the transient composed of a single period. The spectral lines themselves are located at the fundamental and its harmonics. In this case, increasing the number of periods will decrease the computational frequency spacing. The spectral lines are hence more widely separated (in this exercise we note zeros between the lines). For transient 2, the FT of the exponentially decaying oscillating signal shows a resonance around the natural frequency. A periodic signal of such decaying oscillating transients has a spectral line whose envelope tracks the resonance curve – both proportional to $|X(k)|$.

Exercise 3.9

Zooming in (Figure E3.22), the following can be observed: For a fixed sampling interval, changing N, the frequency spacing of the FFT will change, while the maximum frequency, the Nyquist frequency, is fixed.

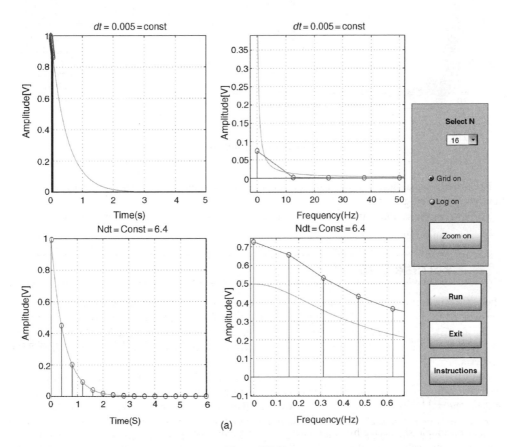

(a)

Figure E3.22

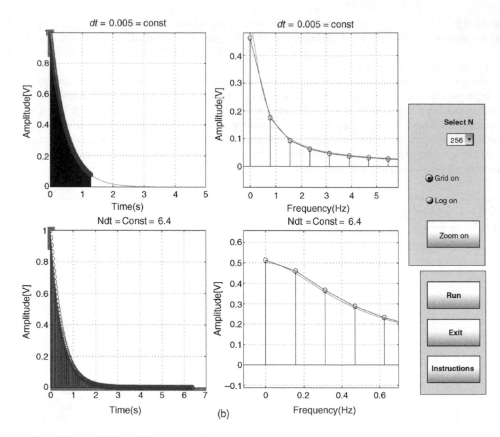

Figure E3.22 (*continued*)

For a fixed signal duration, the sampling interval will change with N. The frequency spacing, is however, constant.

The smaller the sampling interval, the closer the DFT is to the theoretical continuous FT.

Exercise 3.10

The basic transform is the DFT, which may suffice for qualitative representations. Quantitative results, however, depend on the signal class. While the frequency domain pattern may equal that of the DFT, noting the EU used is imperative. Thus we note the different EU for transients (FT, V-sec], periodic (FS, V) and random (PSD, V^2/sec). The PSD will be addressed in detail in Chapter 7, Spectral Analysis.

The one-sided spectral representation may be preferred for application-oriented results, however the comparison with the two sided one needs to be done carefully. There is mainly a factor of 2 for the values of the one-sided one. However, there may be some differences, as noted for the case of the transient signal, where a DC component existed: this has no symmetrical component in the two-sided representation, and its value (unlike the other components) is the same in both presentations.

Exercise 3.11

One possible way of interpreting the sequences $x(n)$ and $X(k)$ is that they are periodic, both with a period of N. The DFT pair are N point sequences, each being *one period* of a periodic discrete sequence. The parameter N controls the time period spanned, and the resolution (frequency spacing) in the spectrum.

It is important to note the type of scales, as obvious patterns may be destroyed (changed) by nonlinear scales.

Exercise 3.12

When an even number of half periods is chosen, no discontinuities exist in the DFT (which consist of N points of a periodically repeating sequence). Spectral lines occur at the corresponding frequency of the signal. When choosing four half sections, the basic N points span two periods, and the spectral line occurs at the index $k = 2$ (third point as k starts from zero), as can be noted using the zoom option. The additional lines repeat (negative frequency at $k = 510$ and at the corresponding locations of the extended sequence). Applying a window widens the spectral representation, with two additional lines being generated around the original ones.

For an odd number of half periods, discontinuities appear in the extended time signal (Figure E3.23).

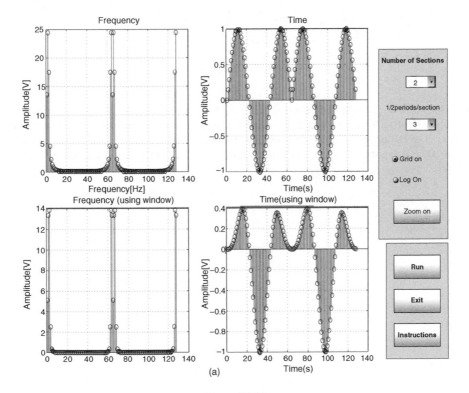

Figure E3.23

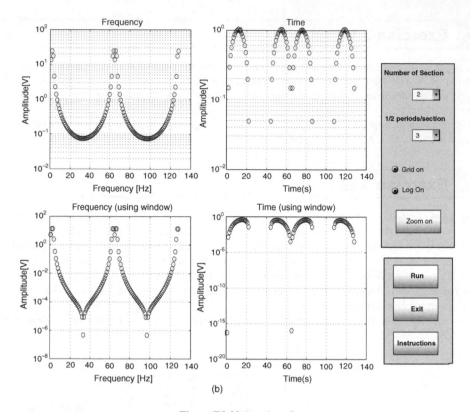

Figure E3.23 (*continued*)

The discontinuities in the time domain result in the spread of the spectrum (as predicted from the Gibbs effect). This is more clearly seen by using log scales, showing how significant this spread is. The advantage of applying a window, to avoid the discontinuity, is evident, again more obvious when using the log scale. (Note the minimum values in the spectra with and without the window!)

4

Linear Systems

Overview

These exercises are intended as refreshers on the topic of linear systems (a prerequisite for this book). Exercise 4.1 shows the analogies between continuous and discrete systems. The specific examples of accelerometers, Exercises 4.2 and 4.3, demonstrate how instrumentation specifications can affect the type of signals to be processed in practice.

The Exercises
Exercise 4.1

Objective

To describe continuous and discrete linear systems via different types of parameters, and show responses to typical excitation signals. The filters (Chapter 5) will then be a specific case of such systems.

Reminder

- For continuous systems:

$$\sum_{k=0}^{N} a_k \frac{d^k}{dt^k} y(t) = \sum_{k=0}^{M} b_k \frac{d^k}{dt^k} x(t)$$

$$A(s) = \prod_{k=1}^{N} (s - z_k)$$

$$B(s) = b_m \prod_{k=1}^{M} (s - p_k)$$

$$H(s) = \frac{B(s)}{A(s)}$$

and the frequency response is

$$H(j\omega) = H(s)\big|_{s=j\omega}$$

- For discrete systems:

$$\sum_{k=0}^{N} a_k y(n-k) = \sum_{k=0}^{M} b_k x(n-k)$$

$$H(z) = b_m \frac{\prod_{k=j}^{M} (z - z_k)}{\prod_{k=1}^{N} (z - p_k)} \qquad \sum_{k=0}^{N} a_k z^{-1} Y(z) = \sum_{k=0}^{N} b_k z^{-k} X(z)$$

and frequency response is

$$H(j\omega) = H(z)\big|_{z=\exp(j\omega\Delta t)}$$

Description

Various systems and their response to specific excitations can be simulated. The two upper plots (Figure E4.1) show a description of a continuous system (left), and a discrete one (left) which is actually

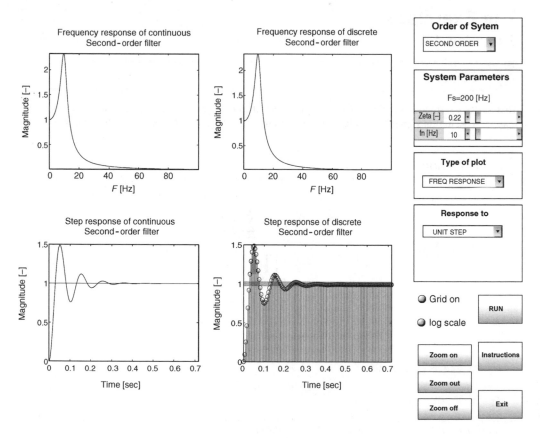

Figure E4.1

a simulation of the continuous one. Two types (first and second order) systems can be chosen, together with varying system parameters for each one. The sampling interval used for the discrete systems is also noted. Systems can be described (upper plots) by either their frequency response, zeros/poles configuration or their impulse responses.

The two lower plots show the system responses (blue) to possible excitations (green), step or harmonic. For the latter case the excitation frequency and the duration (number of cycles spanned) is also controllable. As above, values can controlled via sliders or by typing desired values in the corresponding boxes.

Instructions

Choose a first order system, varying its parameter (time constant tau). See the effect on its various characteristics (FRF, pole/zeros, impulse responses). See the effect on the time response for step and harmonic excitation – also varying the parameters of the latter.

Repeat for second-order systems.

Tasks

- Summarize which functions/data are continuous/discrete.
- For a first-order system (discrete and continuous), summarize the effect of tau on:
  - the pole/zero description
  - the frequency response
  - the step, impulse and sine response.
- Repeat for a second-order system, summarizing the effect of zeta and the natural frequency.
- Choose $fn = 30$, $z = 0.7$. Choose a sine excitation of 30.1 Hz. Repeat for $z = 0.05$ and explain all results.

Exercise 4.2

Objective

To explore the response of an accelerometer to a noisy transient.

Reminder

The transfer function of an accelerometer is that of a second-order low pass system.

$$H(s) = \frac{K}{\omega_0^2} \cdot \frac{1}{\dfrac{s^2}{\omega_0^2} + 2\zeta \dfrac{s}{\omega_0} + 1}$$

with K a gain factor.

Description

An accelerometer, whose natural frequency and sensitivity are controllable (pop-ups under 'Sensor properties'), is excited by a transient with additive noise. The transient's duration as well as the additive noise are controllable (pop-ups under 'Signal properties'). The noise-free transient is shown in the upper left plot, the noisy one in the upper right. Bode plots for the accelerometer are shown in the two middle plots, the gain in decibels. Three scenarios can be chosen via the 'Analysis' pop-up: excitation by a noise-free transient, by the noise only and by a noisy transient. The last case is just a superposition of the first two. The response is shown in the lower left plot, both the excitation and response in the lower right (Figure E4.2).

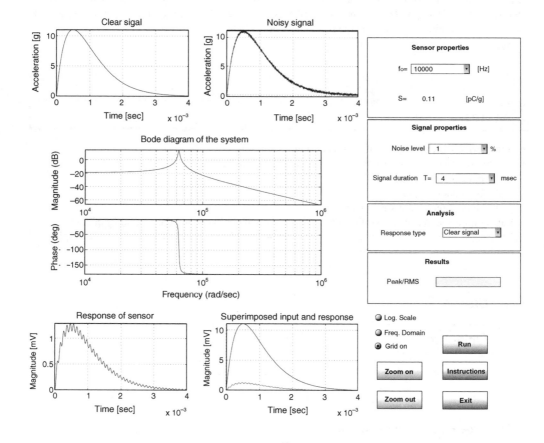

Figure E4.2

For the case of a noisy transient, an additional numerical quantity, peak/RMS is shown in the 'Result' frame: this is the ratio of the peak for the ideal (noise-free) transient response to that of the output RMS with noise excitation only.

Instructions

For a noise-free transient, investigate the effect of changing the sensitivity, natural frequency and the transient duration
   Repeat for a noisy transient (for various noise levels), checking also the resulting peak/RMS value.

Tasks

For noise-free transients determine and explain the sensor's response and that of the superimposed oscillations as a function of the transient duration and the accelerometer's natural frequency.
   Checking the noise output only, explain the effect of the sensor's natural frequency. For a noisy transient, check the peak/RMS value for various natural frequencies and explain the results.

Exercise 4.3

Objective

To compare the performance of two accelerometers, and summarize criteria for choosing one.

Reminder

The transfer function of an accelerometer (force applied to the sensing device per acceleration) is that of a second-order low pass system, with K a calibration factor:

$$H(s) = \frac{K}{\omega_0^2} \frac{1}{\dfrac{s^2}{\omega_0^2} + 2\zeta \dfrac{s}{\omega_0} + 1}$$

For the operating range $\omega < \omega_0$, the gain is inversely proportional to the natural frequency.

Description

This exercise is similar to Exercise 4.2, except that the response of two accelerometers is compared. They differ in sensitivities and natural frequencies.
   The same excitation is applied to both accelerometers, shown in the upper plots (left noise free, right noisy; Figure E4.3). The transient's duration and the additive noise are controllable via the 'Signal properties'. Three types of responses can be computed via the pop-up in the 'Analysis' section: for a noise-free transient, for noise input only and for a noisy transient. The responses of both accelerometers are shown in the middle plots. The two lower plots superimpose the responses and excitations. The peak/RMS values (see Exercise 4.2) are also noted for both accelerometers.

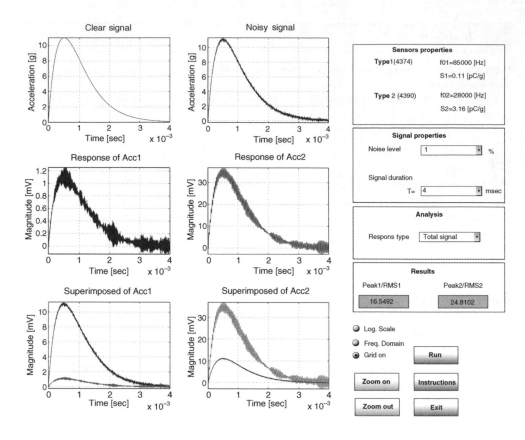

Figure E4.3

Instructions

For the shortest signal duration, compare the outputs of the two accelerometers. The comparison should be made for the response to the noise-free signal, to the noise itself and to a noisy signal. Note the peak/RMS value for both accelerometers. Repeat for the longest duration input signal.

Tasks

- Describe the different outputs of both accelerometers, both for the noise free and the noisy input. Explain the differences.
- Repeat the above for different signal durations, and explain the effect of the signal duration.
- Comment/justify on the different parameters for both accelerometers.
- Summarize the considerations for choosing a 'best' accelerometer, based on their natural frequency.

Solutions and Summaries

Exercise 4.1

Signals are obviously continuous/discrete for continuous/discrete systems respectively. The frequency response is continuous for both type of systems. Zero/poles are in the s or z complex planes. For stable systems, poles are in the left-hand s plane or within the unit circle of the z plane. (Figure E4.4).

For first-order systems, increasing tau results in a narrower bandwidth, the pole in the s plane moves closer to the origin, that of the z plane closer to the unit circle (Figure E4.5).

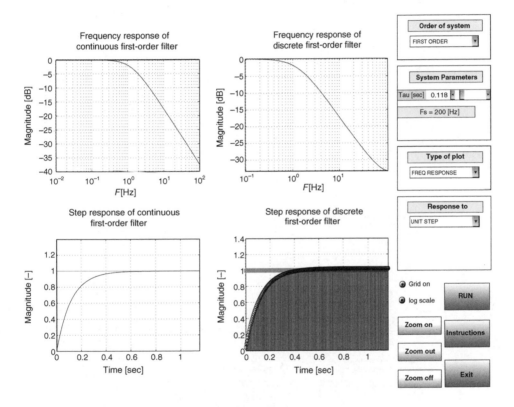

Figure E4.4

For second-order systems, the two parameters fn and zeta determine the properties. Decreasing zeta (say to 0.05), move the complex conjugate poles closer to the imaginary axis (s plane) or unit circle (z plane- Figure E4.6), the frequency response then shows a resonance. In the time domain, this is shown by the magnification of the response when choosing a sine excitation with frequency close to that of resonance. A transient response is also evident, until the harmonic response reaches steady state (Figure E4.7).

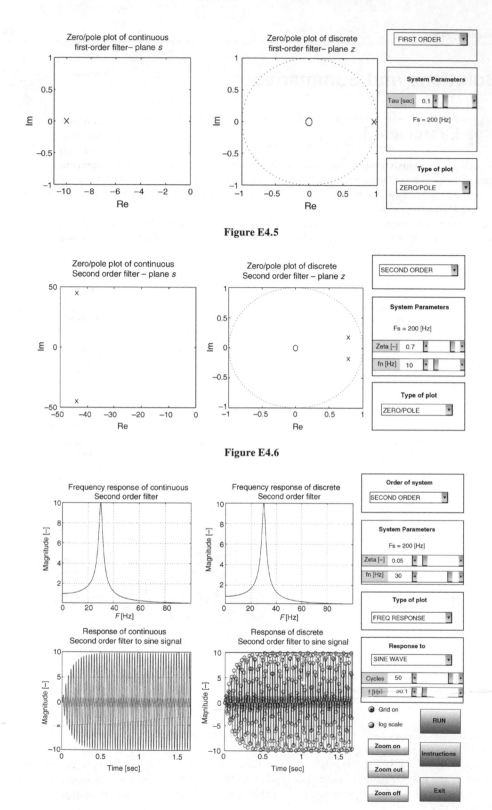

Figure E4.5

Figure E4.6

Figure E4.7

Exercise 4.2

From the noise-free situation, the sensitivity (if controlled independently, say in conjunction with an amplification device) does not affect the dynamic response per se. For a signal duration close to the reciprocal of the resonant frequency (1.2 msec, viz. 1/1000 for 1000 Hz), we note superimposed oscillations at this resonant frequency (Figure E4.8). A higher accelerometer resonance frequency is needed in order to reduce these oscillations.

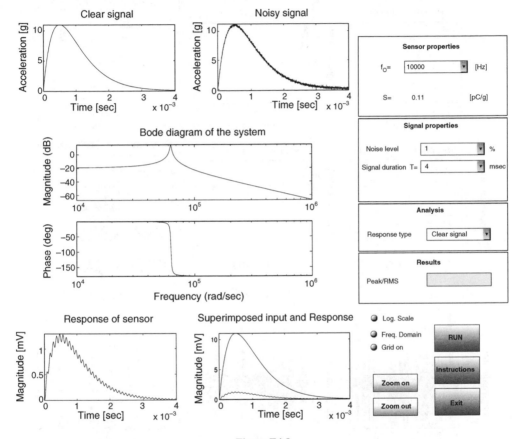

Figure E4.8

The noise response shows these oscillations clearly. By using the zoom options, the approximate noise frequency is seen to be that of the resonant frequency. The peak/RMS value is reduced by a higher resonant frequency. The larger bandwidth results in an increased noise output.

Exercise 4.3

The characteristics of the responses are, of course, those already noted in Exercise 4.2: the shorter excitation will excite oscillations at the accelerometer's natural frequency, unless the natural frequency is high (in relation to 1/signal duration), as seen for type 2 (Figure E4.9). For a longer duration, the oscillations are negligible for both types. A noisy excitation will show significant oscillations for both types, the oscillations being at the natural frequency, 85 and 28 kHz for types 1 and 2 respectively (use of zoom to determine the period is needed).

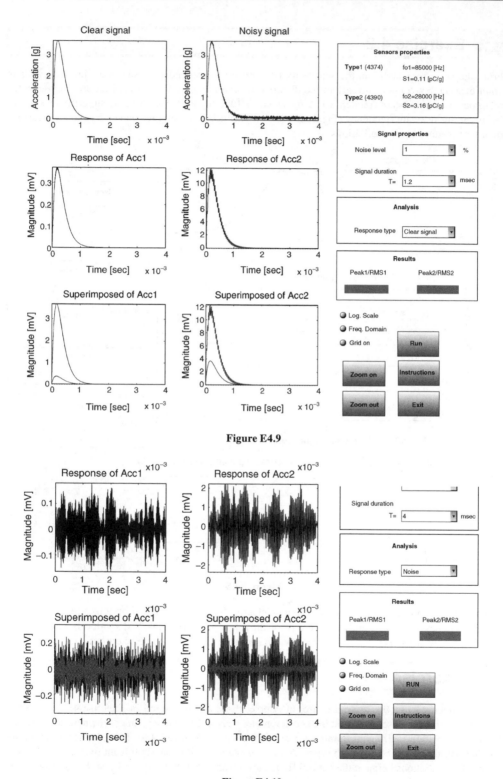

Figure E4.9

Figure E4.10

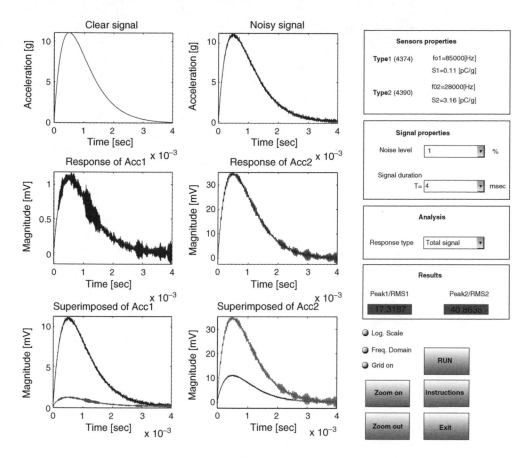

Figure E4.11

To compare the performance of both accelerometers, it is not enough to check the response magnitude. The noise response is much higher for the high frequency sensor, as shown in Figure E4.10. A more objective comparison is the peak/RMS value, which is markedly different for both cases. It is much higher for type 2. Thus in this case, the slower sensor gives a better performance. It has the better balance of a reasonable response to the signal, with higher noise rejection (Figure E4.11).

5

Filters

Overview

In this chapter, the emphasis is on filter *performance*, assuming that the actual filters can be designed and applied with relative ease. The basic smoothing operation of a low pass filter is addressed by Exercise 5.1, while filtering of specific signals is done in Exercise 5.2. The possible need for linear phase filters is demonstrated via Exercise 5.3. A tutorial oriented case of a running average filter is presented in Exercise 5.4.

The Exercises
Exercise 5.1

Objective

To demonstrate some basic aspects of low pass filtering.

Reminder

A low pass filter attenuates high frequency components, beyond a critical frequency, and thus has a sluggish response.

Description

Two types of signals can be chosen: a transient and a step signal. Both clean and noisy (with additive white noise) versions can be processed.

A low pass filter with controllable critical frequency as well as filter order can be applied to the signals. Both the original signals and the filtered ones (for both uncorrupted and noisy cases) are depicted (Figure E5.1).

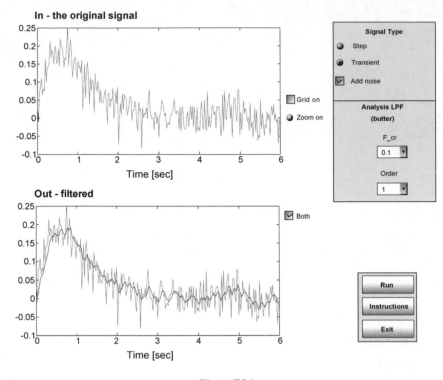

Figure E5.1

Instructions

See the effect of applying the low pass filter to the uncorrupted signal, checking the effect of filter critical frequency and filter order. Repeat for the noisy signals.

Tasks

Summarize the effect of the filter's parameters. With the objective of attenuating the noise, attempt to find the 'best' filter.

Exercise 5.2

Objective

To exercise the basic filtering possibilities: low, high, band and band stop. To introduce linear phase filters.

Reminder

- Low pass filters attenuate high frequency components.
- High pass filters attenuate low frequency components.
- Band pass filters attenuate components outside a frequency band.
- Band stop filters attenuate components inside a frequency band.
- Linear phase filters have constant delay characteristics in the band pass range:

$$\text{delay} = -\frac{d\phi(\omega)}{d\omega}$$

Description

Two types of signals are available for filtering: there is an oscillating one, of exponentially decaying amplitude, and a noisy half sine transient. An externally supplied signal can also be filtered (use the Instruction button for details). All four basic filters (low pass, high pass, band pass and band stop) can be chosen. The critical frequencies are controllable via cursors and the filter orders via a pop-up menu. The right plots (Figure E5.2) show the original (upper right plot) and filtered (lower right plot) signals. Signal spectra as well as the filter's frequency responses are shown in the left plots. To

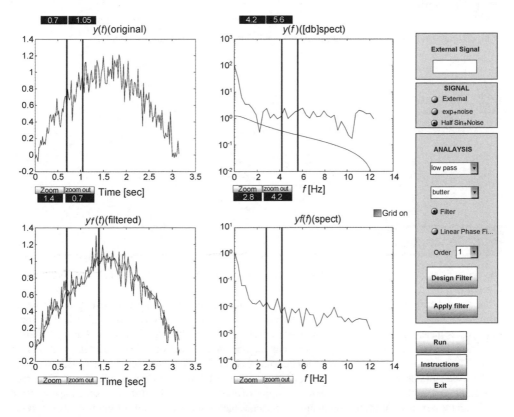

Figure E5.2

perform a filtering operation, a signal is chosen, and the 'Run' button applied. The signals and their spectra will be shown. Next the filter type and order, and also critical frequencies (via cursors) are chosen, and the 'Design filter' button used. The filter's frequency response is then shown in the upper right plot. The 'Apply' button is used to perform filtering, the result appearing in the lower plots. Options are either a classic infinite impulse response (IIR) filter, or a linear phase filter, specifically a zero phase one.

Instructions/Tasks

Choose the noisy half sine signal. Design a low pass filter, check the effect of decreasing the critical frequency (using a high order filter). To be investigated are the remaining noise level and peak location after filtering. Compare the use of a nonlinear and linear phase filter. Compare with Exercise 5.1.

Repeat for a high pass filter.

Choose the decaying oscillating signal. Try to explain its composition by (1) isolating the first spectral peak by a low pass filter, and (2) band pass filtering around the first peak, and next around the second peak.

Exercise 5.3

Objective

To investigate the advantage of using linear phase filters when measuring transients.

Reminder

Linear phase filters have constant delay characteristics in the band pass range:

$$\text{delay} = -\frac{d\phi(\omega)}{d\omega}$$

Description

Two types of signals can be chosen: step and two transients (Figure E5.3). The first case is relatively self-evident.

The second case simulates a situation where a two-channel monitoring scheme detects two transient trains, who are synchronous with the same frequency (for example, body acceleration and cylinder pressure, as measured on a combustion engine). The two channels detect transients having different frequency regions, and filters are applied to them, intended to remove noise. The time delays introduced by these filters may affect the relative time occurrences between the two detected transients, modifying the detected time delay observed. A block diagram (Figure E5.4) shows this configuration. Choosing the signals and applying 'Run' will show the signals (Figure E5.5). The filtered results appear in the middle and lower plots. With the zoom option it is possible to inspect sections of interest of the filtered signals.

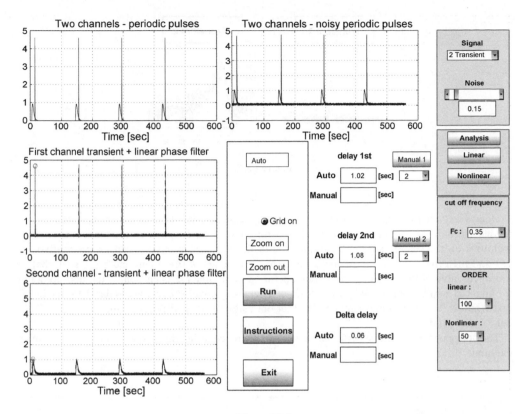

Figure E5.3

For the two-transient case, an option of additive noise exists. The noise-free and noisy signal are shown in the left and right upper plots. Two types of low pass filter can be chosen, with linear and nonlinear phase characteristics respectively. Various critical frequencies and filter orders can be chosen.

The peaks of the original and filtered signals are denoted by a stem and circle at the origin of the signals (zoom is advised).

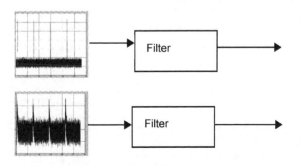

Figure E5.4

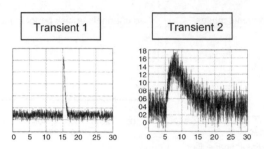

Figure E5.5

Delays between the original and filtered signals can then be determined in two modes. The modes are manual or automatic, to be chosen for each filtered signal. In the manual mode, the locations on the screen can be determined. With the left mouse button, two locations, first the left and then the right peak, are clicked on. This is repeated k times consecutively, such that an averaged value of this possibly inaccurate operation is obtained. k is chosen by the pop-up below the manual pop-up button. When the k set of clicks are done, the average delay appears in the text window. When the delays are obtained for both signals, their difference appears in the 'Delta delay' text window.

The above operation can be performed automatically by choosing the automatic instead of the manual mode.

Instructions

Choose the step signal. Check the effect of applying a linear as compared to a nonlinear phase characteristic filter.

For the two-transient case, choose zero noise, a critical frequency of 0.05 and orders of 100 and 50. Zoom on the first peaks of filtered signals (easily seen by the stem and circles on top). Using the manual mode (with $N = 2$ averages), determine the delays introduced by the filters. Compare with the results in the automatic mode.

Repeat with maximum additive noise.

Task

Summarize and explain when the choice of linear phase characteristics is desirable.

Exercise 5.4

Objective

To explore the characteristics and performance of moving average filters.

Reminder

The frequency response of the moving average filters is

$$|H(j\omega)| = \frac{1}{M} \frac{\sin(\omega M \Delta t / 2)}{\sin(\omega \Delta t / 2)}$$

Description

A moving average filter can be applied to various signal types, chosen via the signal type source. Dat1 is a combination of harmonic and decaying oscillating signals, dat2 a single decaying oscillating signal (of different frequency) with additive noise. The option 'create a signal' generates a combination of a sine signal and a decaying cosine signal, with optional additive noise. The amplitude and frequencies, as well as the noise level, are controllable. An external signal option is also available. The order of the moving average filter is controllable in the range of 1–40. The upper left plot (Figure E5.6) is that of the signal to be filtered, with the noise free signal in red, the noisy in blue. The right upper plot is of the

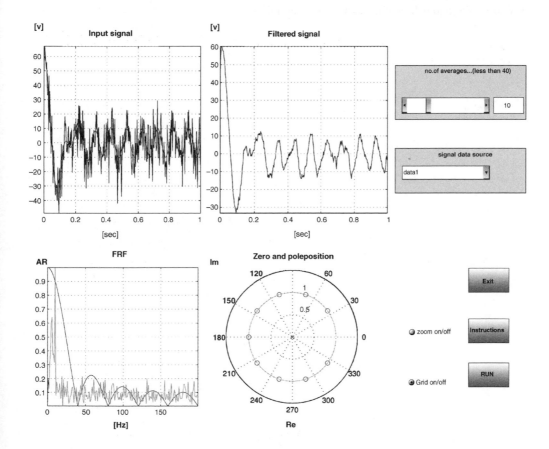

Figure E5.6

filtered signal. The lower left plot shows the frequency response function (FRF) of the filter, superimposed on the signal's spectrum, while the lower right plot shows the poles/zeros of the filter.

Instructions

Choose dat2, attempt to attenuate the main signal components. Repeat for dat1. Create a signal with these parameters: sine and cosine frequencies of 15 and 35, amplitudes of 5 and 3. Filter components in the noisy and noise-free situations.

Tasks

- Check the effect of the filter's order on the lobe maxima and zeros.
- Check the possibilities of cancelling components by aligning corresponding peaks with zeros of the filter.
- Check the attenuation of components in the time and frequency domains.
- Check the effect of the various filter orders chosen on the output noise.
- Explain the zero/pole configuration for various orders.

Solutions and Summaries

For the noise-free case, the filtered signal is delayed and 'smeared' (Figure E5.7). Both effects become more pronounced as the critical frequency decreases and the filter order increases (Figure E5.8).

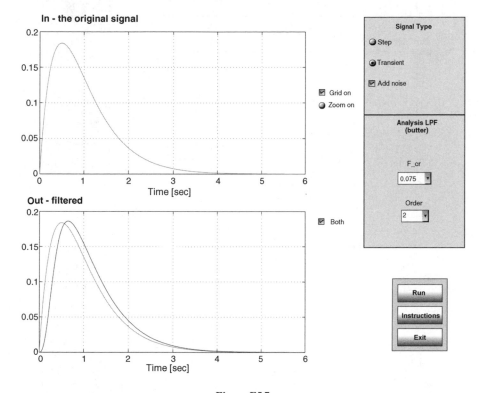

Figure E5.7

For a noisy signal (Figure E5.9) the two parameters have to some extent similar effects. Increasing the order and decreasing the critical frequency, while increasing the noise's attenuation, will also delay and distort the original signal.

It is not trivial finding the 'best' filter. What is needed is a filter with band pass characteristics matched to the signal's frequency content, and band stop matched to that of the noise. If the frequency characteristics of these two components overlap, then a 'best' filter can only be determined if a criterion for 'best' exists. One possible criterion could be to maximize the output signal to noise ratio, but this would need a priori knowledge of the signal and noise characteristics. *Hence the seemingly simple question of choosing the best filter parameters has no trivial answer.*

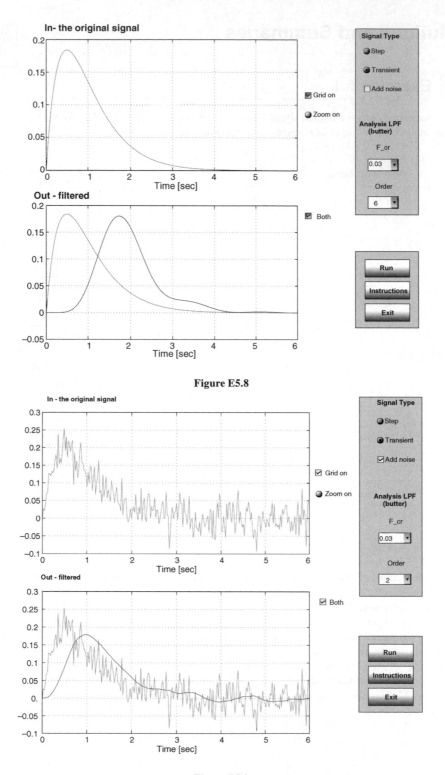

Figure E5.8

Figure E5.9

Exercise 5.2

For the noisy half sine case, we note results similar to those of the previous exercise (Figure E5.10). Decreasing the critical frequency gives results as shown in Figure E5.11.

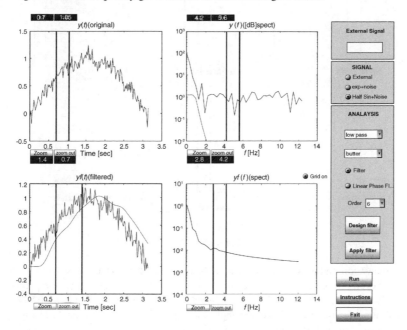

Figure E5-10

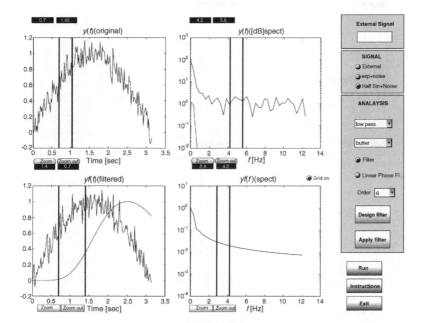

Figure E5.11

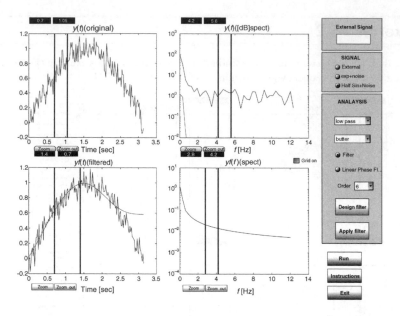

Figure E5.12

The noise attenuation and delays increase with a decrease in a filter's band pass bandwidth. A linear phase filter, however, can be used to have a controlled delay. For the linear phase equaling zero, no delay is introduced, and the output peak is now aligned with that of the input (Figure E5.12).

For the oscillating decaying signal, we first apply a low pass filter to the first peak. The filtered signal has the shape of the original envelope (Figure E5.13). Applying a band pass filter around the second

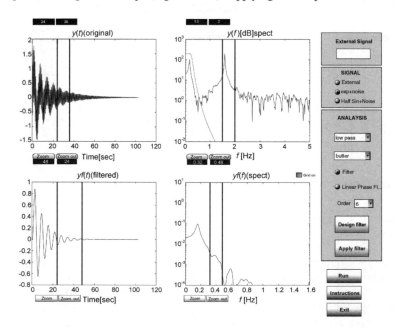

Figure E5.13

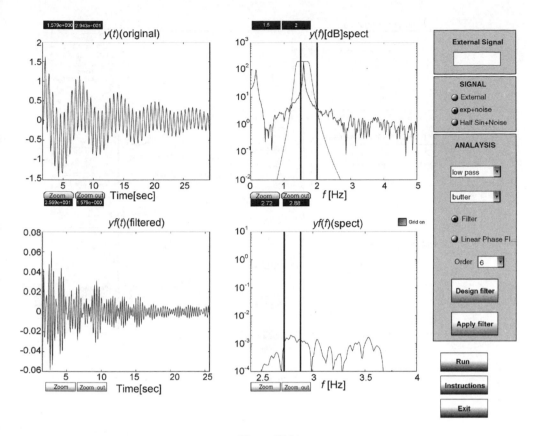

Figure E5.14

peak gives results as shown in Figure E5.14 and only the high frequency part now results. Thus the signal is mainly composed of the sum of two decaying components, with low and high frequencies respectively.

Exercise 5.3

Results for the step signal nonlinear phase characteristics are shown in Figure E5.15 and for linear phase characteristics in Figure E5.16. For such cases, involving a sharp transient, the advantage of the linear phase response is evident.

For the two-transient case, choosing nonlinear characteristics, and next applying the manual and then the automatic mode, results in the two signals being shifted by delays, the difference in which is between 40 and 60 msec (Figure E5.17). With linear phase characteristics (Figure E5.18), the two delays are practically the same.

For the noisy case, the peaks are not well defined, and it is not easy to determine exact delays (Figure E5.19).

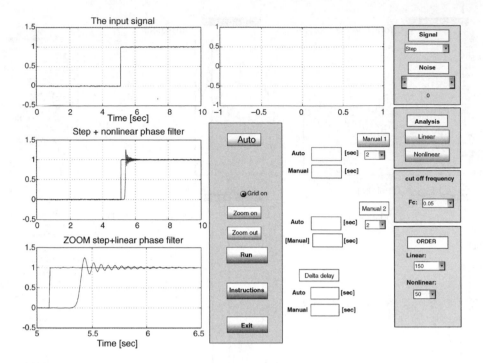

Figure E5.15

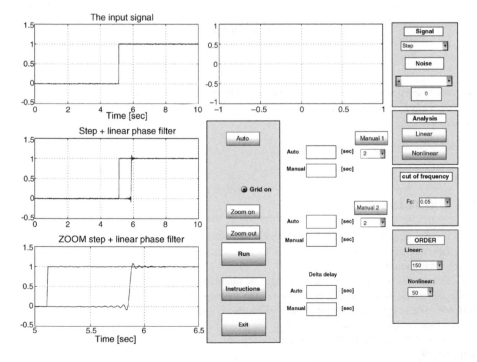

Figure E5.16

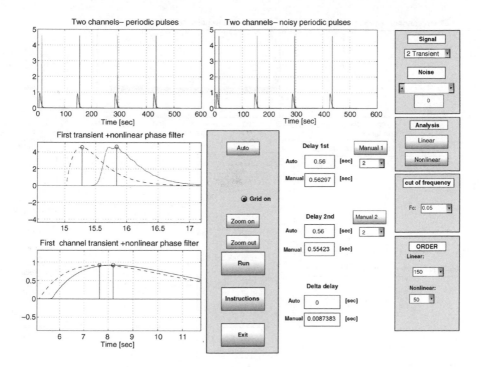

Figure E5.17

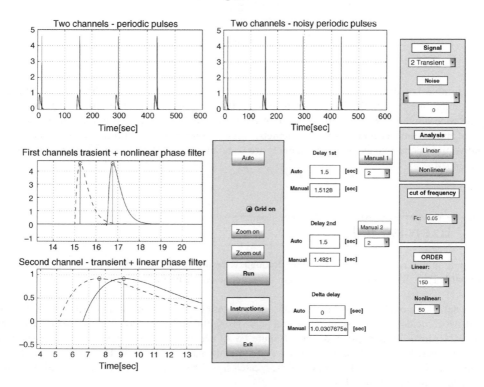

Figure E5.18

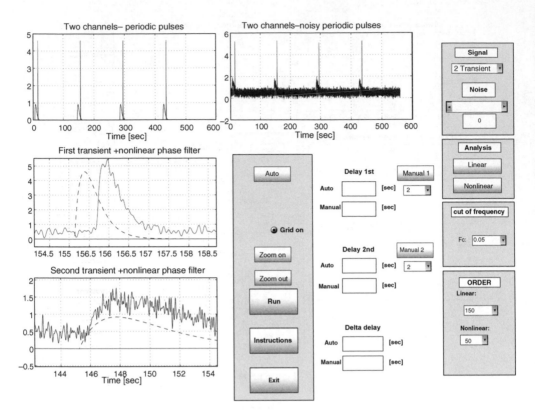

Figure E5.19

To summarize, phase distortion due to nonlinear phase characteristics can have a marked influence on the filtered signals. This seem to be typical in the case of transients with sharp transitions in the time domain, as shown for the case of the step signal. An interesting and practical case is that of measurements involving two channels, when different type of filters are used in each.. If relative time occurrences between channels are to be preserved, then linear phase filters are recommended.

Exercise 5.4

The pole and zero are cancelled at zero frequency, resulting in low pass characteristics. The additional zeros correspond to the zero gain at frequencies located at midpoints between secondary lobes.

For dat2, it is possible to choose the number of averages (26 in this case), such that the first filter zero, at 30 Hz, coincides with the signal's spectral peak. Using the zoom options, an almost exact matching is possible. Inspecting the filtered signal (Figure E5.20 upper right) shows that the oscillations seem to be attenuated completely after 50 msec. A similar attenuation for the 10 Hz oscillation of dat1 is achieved with 40 averages. The option of the created signal generates a combination of dat1 and dat2, with details available under 'Instructions'.

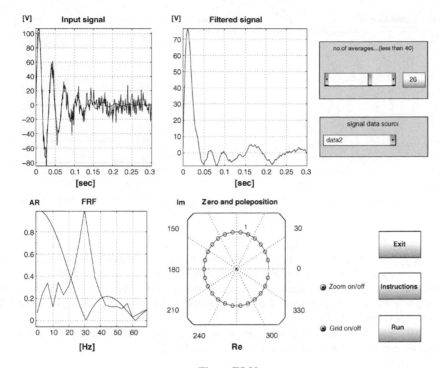

Figure E5.20

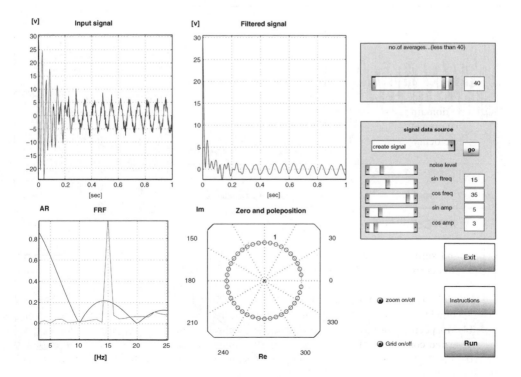

Figure E5.21

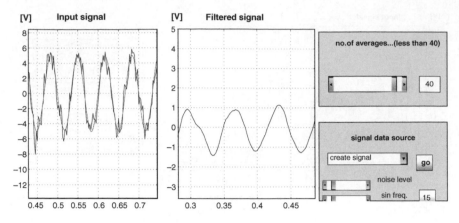

Figure E5.22

After generating the required signal, various attenuations can be achieved by choosing the number of averages. For the required signal parameter values, choosing 40 averages will align the 15 Hz component with the second lobe of the filter, with an attenuation to 20% (Figure E5.21). This can also be noted in the filtered signal (upper right). Checking say at 0.4 sec, we notice an amplitude of 1 V, as compared to 5 V (Figure E5.22).

With 26 averages, the 15 Hz component can be cancelled. With additive noise, however, the filter only attenuates part of the noise (see lower left plot, Figure E5.23).

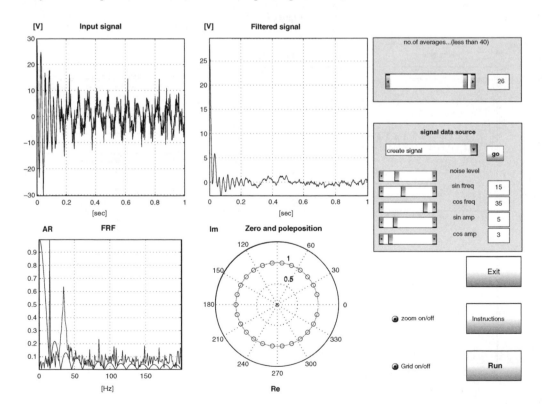

Figure E5.23

6

Time Domain Averaging (TDA)

Overview

The powerful capability of extracting a periodic signal from interfering disturbances, is demonstrated by Exercise 6.1. The dynamic display enables us to judge the performance by visual inspection. The averaging process, viewed as a filter, is addressed by Exercises 6.2 and 6.3, the latter giving an intuitive understanding of how harmonic disturbances are rejected. The effect of jitter is demonstrated by Exercise 6.4.

The Exercises
Exercise 6.1

Objective

To explore the signal extraction capabilities of TDA.

Reminder

- TDA will extract the periodic component, having a line spectrum.
- Random noise will be attenuated as

$$\frac{1}{\sqrt{N}} e_{RMS}$$

- The attenuation of harmonic interference depends on its frequency relative to that of the extracted periodic signal.

Description

Three types of signals can be chosen, as well as two types of additive noises. For random noise, only the level can be controlled; for harmonic noise, both the level and frequency. The number of averages can be chosen, and for the final display the length of the section averaged (chosen as the number of 'Periods in group'). Displayed in Figure E6.1 are the composite signal (top), the composite and averaged one (middle), and the spectral description of both (bottom).

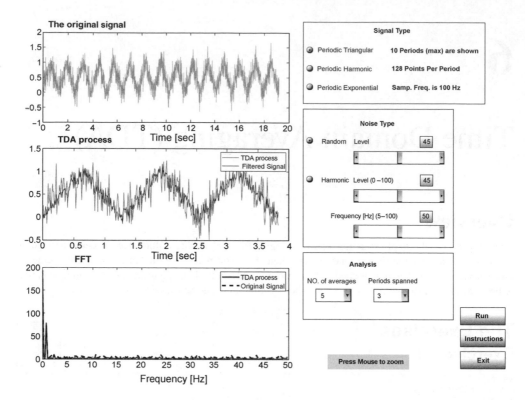

Figure E6.1

Running the program results in a dynamic display, and the evolution of the averaging process can be followed very effectively by visual inspection.

Instructions

Choose a noisy sine wave, see the effect of the number of averages on the additive random noise. Repeat for a sine additive interference, investigating also the effect of its frequency.

Repeat for triangular and decaying oscillation periodic signals.

Tasks

- To demonstrate/note the effect of averaging in the time and frequency domains.
- To compare the above for all three types of signals.
- To note the difference between the attenuation of additive random or harmonic disturbances.

Exercise 6.2

Objective

To explore the rejection of harmonic interferences via TDA.

Reminder

- The TDA's transfer function has a shape consisting of main and secondary lobes.
- The local envelope (the maxima) has the form

$$H_{\max}(k) = [N \sin (\pi f / f_{\mathrm{p}})]^{-1}$$

and attenuation of the interference decreases asymptotically as its frequency is more distant from that of the extracted signal.
- The actual attenuation value oscillates as the frequency is changed.

Description

A periodic signal is given, consisting basically of decaying oscillations. Additive noise of harmonic form can be added, with the frequency and amplitude controlled, as well as random noise, with the RMS controlled. The number of averages can be changed. Shown in Figure E6.2 are four periods of the

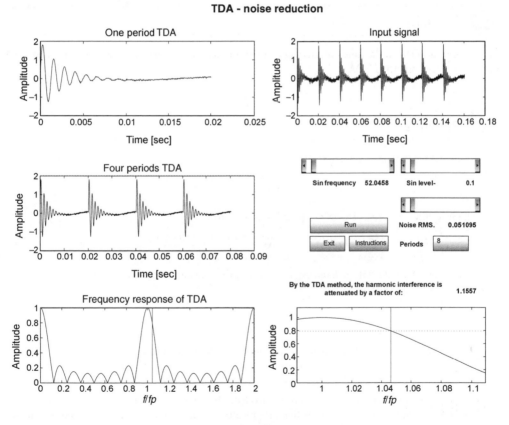

Figure E6.2

TDA extracted signal, as well as the FRF of the TDA process. The frequency of the additive harmonic noise is seen in the FRF, and also in an extended zoomed view (lower right plot).

Instructions/Tasks

Start with zero additive disturbance, then add a harmonic one, and see the effect of the number of averages as a function of frequency of the disturbance. Repeat with random disturbance only and finally with both kind of disturbances.

Some of the following instructions necessitate the use of *exact* sine frequencies. To enable this option, the program should be run from start, and the Sin frequency slider moved *only* by the horizontal scrollers. Good, but somewhat approximate, results will, however, be achieved via the easier option of moving the slider directly.

For harmonic disturbance, see the combined effect of disturbance amplitude and frequency for different number of averagings, investigating the FRF of the TDA. Check for arbitrary frequencies, and then specifically for frequencies of 447.75 and 435.25 Hz respectively.

Repeat Exercise 6.1, find and explain the case of harmonic disturbance, whereby the disturbance seems to grow and decrease alternately during the TDA process.

Exercise 6.3

Objective

To interpret the TDA in the time and frequency domain, and understand the difference between ideal and exponential forgetting TDA.

Reminder

The ideal TDA can be computed recursively by

$$y_r(n\Delta t) = y_{r-1}(n\Delta t) + \frac{x_r(n\Delta t) - y_{r-1}(n\Delta t)}{r}$$

with an 'infinite memory'. However, using the procedure

$$y_r(n\Delta t) = y_{r-1}(n\Delta t) + \frac{x_r(n\Delta t) - y_{r-1}(n\Delta t)}{N_1}$$

the recursive formula with a fixed N_1 results in an exponentially decaying memory, with an approximate time constant of $N_1\Delta t$.

Description

The number of periods to be averaged can be controlled. Shown in Figure E6.3 is the time signal of the averaged signal, whereby each period has the form resulting from averaging the corresponding number of averages. As an example, for 13 periods, we would see on the upper graph between 55 and 65 sec, the result of averaging 10 periods. Thus the evolution of the extracted signal as the number of averages increases can be followed.

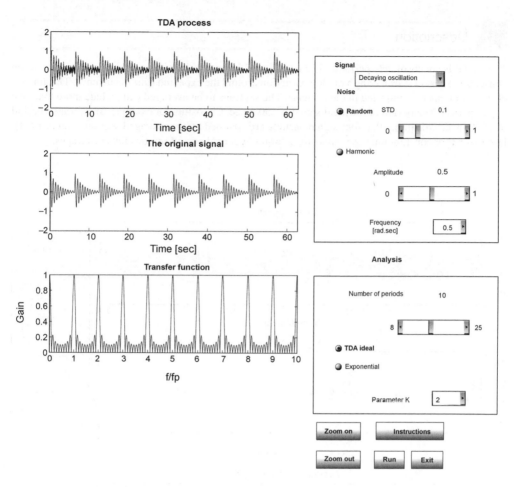

Figure E6.3

Instructions/Tasks

Choose a signal, and add noise. Apply ideal TDA to check the improvement in the signal to noise ratio SNR). Repeat for various numbers of periods averaged, investigating the resultant SNR to the FRF shape.

Compare the results with those obtained with exponential averaging.

Add harmonic noise as follows frequency 4.5, amplitude 8, number of periods averaged 10. Compare ideal TDA with exponential averaging, choosing, $K = 4$.

Exercise 6.4

Objective

To see the effect of signal jitter on TDA.

Description

The signal is not absolutely periodic. The term 'jitter' indicates slight fluctuations in the period, each consecutive period start having a random variation around the expected time occurrence. The amount of this jitter can be controlled in the exercise. The sections to be averaged can include more than one signal period. The period of the ideal signal is 250 msec. Choosing, for example, an averaged period of 1500 msec will imply that four signal periods are spanned by the averaged section (Figure E6.4). For a signal of constant duration, choosing a larger averaged period implies fewer averages.

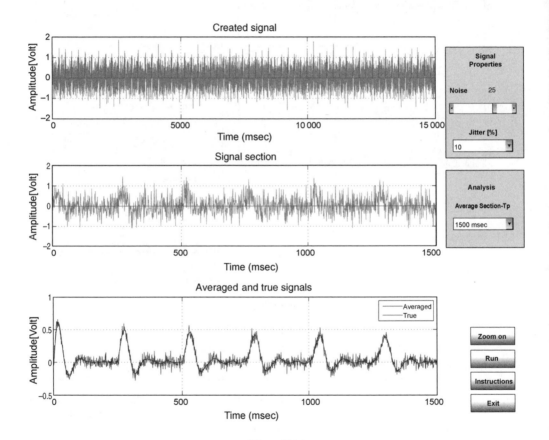

Figure E6.4

Instructions/Tasks

Choose the section to be averaged as 2000 msec, spanning eight periods. For a noise figure of 12, investigate the effect of jitter (including the case of zero jitter).

Repeat for spanned sections of other durations.

Solutions and Summaries

Exercise 6.1

TDA is very effective for the attenuation of additive random noise. This is easily seen for the case of harmonic and triangular signals. Choosing the exponentially decaying oscillations (Figure E6.5), three-period section, with a random noise level of 3, visual inspection of the developing average is almost striking, if seen in both the time and frequency domains. The spectral lines extend beyond 14 Hz, and can only be seen by the averaged (and not raw) signal. The diminishing return for a increasing number of averages seems to be compatible with noise reduction being proportional to $1/sqrt(N)$.

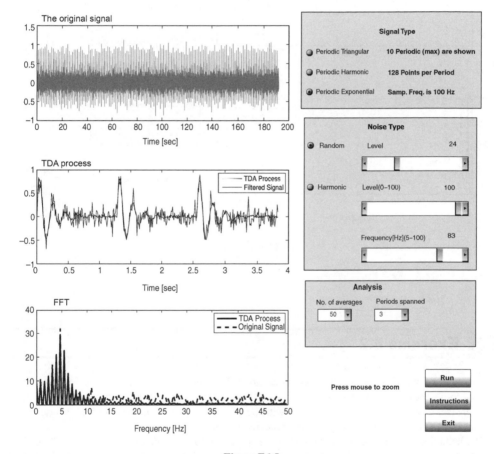

Figure E6.5

For harmonic additive disturbance (Figure E6.6), the effectiveness of TDA depends significantly on its frequency; see, for example, a signal with $f = 83$ Hz and applying 10 averages. As can be easily noted on the spectral plot, for the case of an harmonic noise, the noise reduction can also be asymptotic, i.e via temporary increases/decreases.

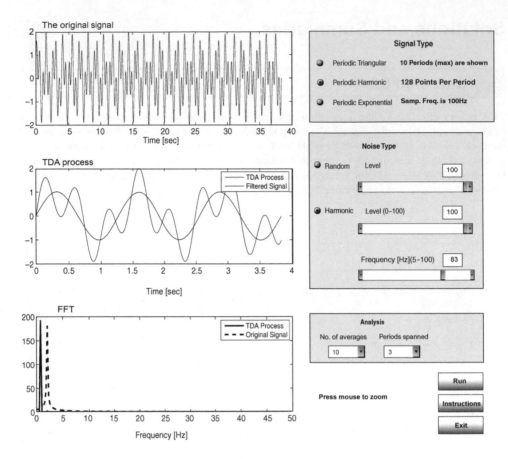

Figure E6.6

The number of averages controls the noise attenuation. Whereas for random noise the decrease with N is monotonic, the decrease of the harmonic disturbance is only asymptotically monotonic. Oscillations, i.e. increases as well as decreases, may occur with the increase of N, and the behavior is highly dependent on the ratio of the signal frequency to that of the disturbance.

The behavior for a harmonic disturbance is easily understood in conjunction with the TDA's FRF. For the frequency of 447.75, the disturbance frequency coincides approximately with the FRF peak, and hence is not attenuated (Figure E6.7). For 435.25 it coincides approximately with a zero, and hence is cancelled out, independent of the disturbance's amplitude (Figure E6.8). (Note these numbers are only approximate, hence the slight dependence on N for the first case and on the amplitude for the second.)

The shape of the FRF is that of a comb filter, with main lobes at integer multiples of the basic frequency (the reciprocal of the averaged period). The filter's basic shape is that of a moving average filter, as also seen before via Exercise 5.4.

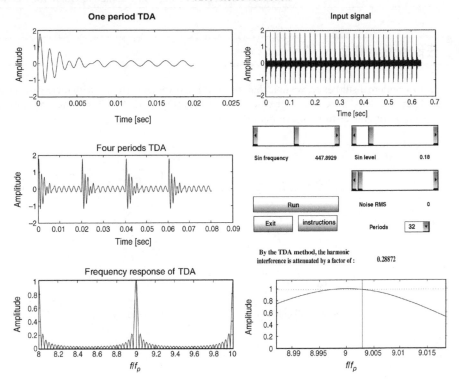

Figure E6.7

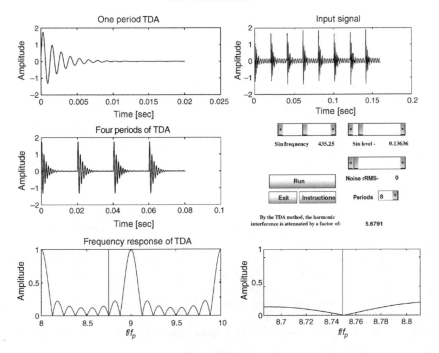

Figure E6.8

Exercise 6.3

Starting with a random disturbance, we note that the SNR improves consistently with increasing number of averaged periods. See, for example, the case of 14 periods (Figure E6.9). This is even more obvious for the case of a harmonic disturbance (Figure E6.10).

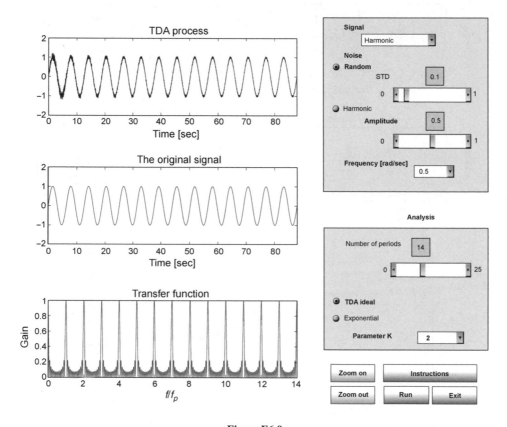

Figure E6.9

For exponential TDA there is an initial improvement which then tends to stay constant. As an example, compare the previous result with that shown in Figure E6.11. This shows a harmonic disturbance case, spanning 14 periods with $N_1 = 10$.

The FRFs of both procedures are different. For ideal TDA we note a comb filter with main and secondary lobes. For exponential TDA, the lobes are similar to a first-order low pass system (say an RC system). The effect may be clearer for a square signal or one of decaying oscillations.

As noted, the difference between the two types of TDA was already quite obvious for the specific case of harmonic disturbance investigated.

The final conclusion is that exponential TDA has characteristics of an exponentially decaying memory.

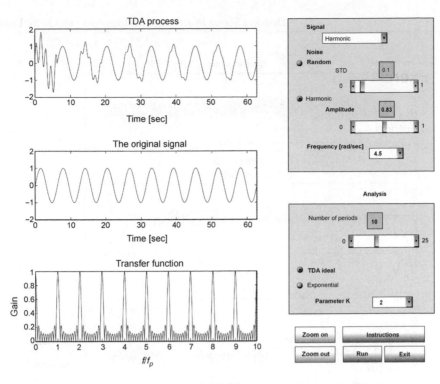

Figure E6.10

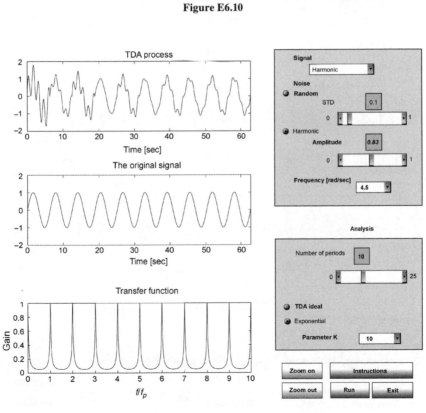

Figure E6.11

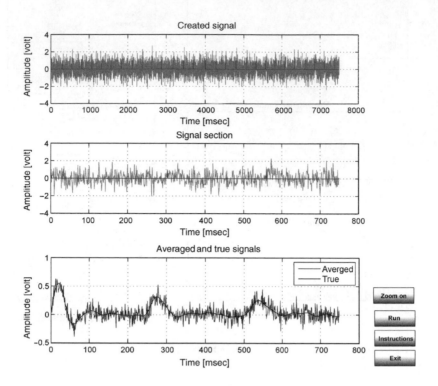

Figure E6.12

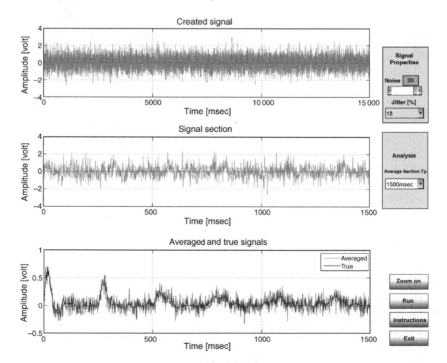

Figure E6.13

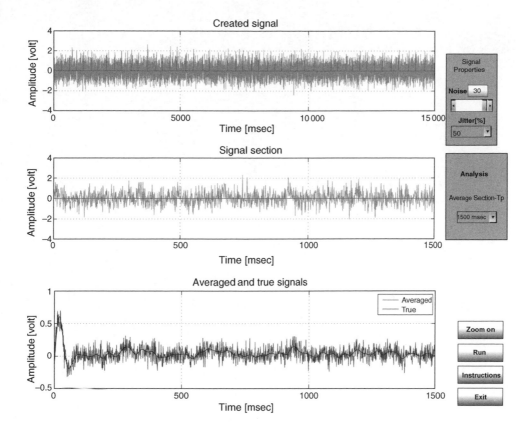

Figure E6.14

Exercise 6.4

The effect of jitter is to attenuate averaged signal sections more distant from the start of the averaged signal. (For zero jitter the result is of course ideal.) Hence jitter is more easily recognized if the number of periods in the averaged sections is higher. This is seen in Figures E6.12 and E6.13 showing three and six sections respectively. The effect of jitter is more pronounced (Figure E6.13)) for signal sections further from the origin. Fewer averages can be performed for larger sections (for fixed duration signals). Fewer sections will be averaged, resulting in a larger residual noise. This is more pronounced for larger jitter and larger N (number of averages), and in extreme cases can destroy the extracted signal shape (Figure E6.14).

We note, as a final conclusion, that signal averages which span more than single signal periods are sensitive to jitter, and can indicate a possibility of jitter.

7

Spectral Analysis

Overview

This chapter concentrates on *applying* spectral analysis. The importance of engineering units is first addressed (Exercises 7.1a and b). This is followed by exercises concerned with leakage, its control via windows and the separation capabilities for closely spaced spectral components (Exercises 7.3–7.5). Errors and their control, for the case of random signals, are addressed by Exercises 7.6–7.8). Spectral analysis via the correlation function is shown by Exercise 7.9, a must for understanding the detection of delays (Chapter 14).

The Exercises
Exercise 7.1(a)

> ### Objective

To use the FFT algorithm to compute a spectrum with EU (engineering units).

> ### Reminder

- The DFT of an N point sequence $x(n)$ is

$$X(k) = \sum_{n=0}^{N-1} x(n) \exp\left(-j\frac{2\pi}{N}nk\right)$$

with

$$X(-k) = X(N-k)$$

- Negative frequencies are associated with $N > k > N/2$ for the two sided transform $X(k)$. A translation of $N/2$ points can give the more used symmetrical representation, with $k = 0$ the zero frequency. A one-sided representation, from 0 to $N/2$, may be used, with an appropriate change of magnitude – a factor of 2 except at end points 0 and $N/2$.

- $X(k)$ being complex, symmetries occur for real signals $x(n)$. Representations of magnitude, real, imaginary and complex displays are possible.
- Engineering units will differ according to signal types:

Aperiodic:	DFT	[V-sec]
Periodic:	FS	[V]
Random:	PSD	[V²/Hz]

Description

This is already described, as Exercise 3.10

Instructions

Exercise, as a reminder, the EU used for the three cases: transient, periodic and random.

Task

Given a sequence $x(n)$, $n = 0,1,2... N - 1$, units in volts, and sampling interval Δt: write a Matlab program to compute a one-sided spectral presentation (with EU) for the cases of the three signal types, transient, periodic, random.

Exercise 7.1(b)

Objective

To note the possibility of different spectral PSD based units, and some signal-dependent behavior.

Reminder

The PSD is

$$S = \frac{\Delta t}{N}|X(k)|^2 \ \ [\mathrm{V}^2 / \mathrm{Hz}]$$

where $X(k)$ is the FFT. The power in each bin (between two spectral lines) is

$$P_{\mathrm{bin}} = S \, \Delta f$$

with

$$\Delta f = \frac{1}{N \, \Delta t}$$

Description

Random or deterministic continuous signals can be chosen, with the option of varying N, the number of data samples. For random signals, the PSD or the power in one frequency bin is computed (and compared to the theoretical result). For deterministic signals, periodic signals can be chosen, N controlled, but an integer number of periods is always dictated.

The computed as well as theoretical spectral values are noted. For random signals, the theoretical values refer to the region of approximately constant PSD.

Zooming into a desired frequency region, a desired range has to be typed in by the user.

The two lower plots show the FFT and this value divided by N, to be used for an FS decomposition (Figure E7.1).

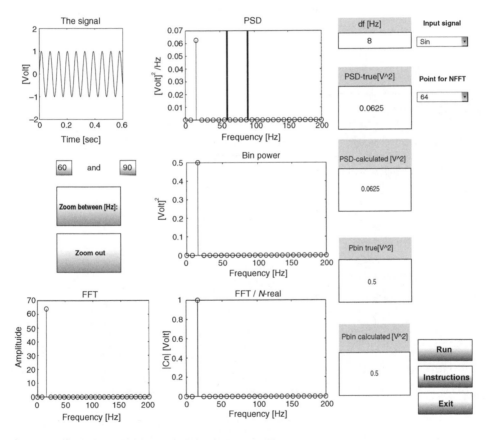

Figure E7.1

Instructions/Tasks

Choose a random signal. Check the effect of changing N both on the PSD and the power in one bin. Check this for various spectral regions.

Repeat for the case of a harmonic signal. Check the effect of changing N both on the PSD and the power in one bin. Check the theoretical results for the PSD, power in one bin and the FS computed by dividing the FFT by N.

Repeat for a square signal.

Also to be checked is the position on the frequency scale as related to the number of periods chosen. Finally the computational result of the FFT and FFT/N, also shown, is to be explained.

Exercise 7.2

Objective

To investigate the existence of leakage and the effect of windowing a harmonic signal.

Reminder

Leakage in spectral analysis is manifested by the occurrence of spectral lines spreading from the region corresponding to the actual frequency of the physical signal.

Applying a window will modify the leakage:

$$x'(n) = w(n)x(n)$$

$$X'(k) = W(k) \otimes X(k)$$

where

$$x(n) \leftrightarrow X(k)$$

$$w(k) \leftrightarrow W(k)$$

Description

The upper plots (Figure E7.2) show the harmonic time signal (left), and the tail end of this signal (right plot). The frequency can be controlled from 1.9531 to 2.0508 Hz via the parameter K, which defines the frequency as a factor of $(20 + K)$ multiplying Δf, the frequency spacing of the FFT. Changing the frequency modifies the number of periods spanned by the time signal, and the red line in the upper right plot facilitates the checking of the periods involved.

The lower plot shows the (absolute) FFT, with a red line showing the true frequency of the signal. Various windows can be chosen for the analysis.

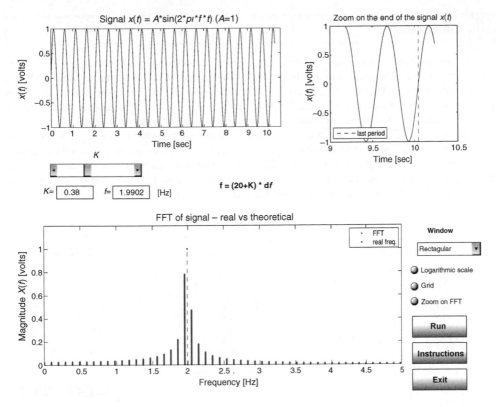

Figure E7.2

Instructions/Tasks

The parameter controlling the leakage is K, the noninteger part of the number of periods spanned by the signal. It is desired to see the scallop loss (the attenuation of the maximum DFT), and the leakage as a function of K. Next, the same is to be repeated after the application of some of the available windows. Whenever desirable, use the log scale.

Exercise 7.3

Objective

To check the existence of leakage and the effect of windowing on two close harmonic signals. To check when two such components can be resolved by the analysis.

Description

The exercise generates the sum of two harmonic signals. The amplitude and frequency of one component are constant. The amplitude and frequency of the second component can be controlled. The frequency f_2 can be varied from $20\Delta f$ to $20 + m\Delta f$ via the parameter m. The control is via sliders, or alternatively by typing a desired value in the boxes above them (the slider position will then change accordingly). Either a rectangular or Hanning window can be applied to the total signal.

Shown in Figure E7.3 are the two components (upper plots), the windowed result (middle plots) and the FFT (lower plot). The displayed frequency resolution can be chosen to be smaller than the computational resolution Δf by a factor of 1–5, controllable by the lower slider (or by typing in a desired value). Then when zooming in, this can enable a more accurate checking of the separation between the two spectral lines. The theoretical spectrum is shown in red, the computed in blue.

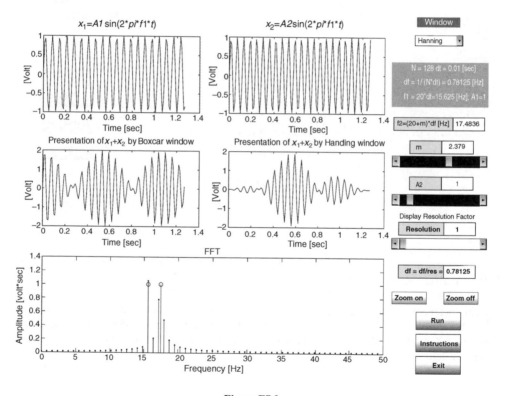

Figure E7.3

Instructions/Tasks

Start with a rectangular window, choose $m = 2$, an integer number of periods for both signals (signal 1 always has such an integer number). Both spectral lines should appear clearly in the frequency domain. Experiment with the change of m to a noninteger number while also changing the amplitude of the second component. The objective is to see when the two signals can be seen as being separated, while comparing the effect of applying a rectangular or a Hanning window.

Exercise 7.4

Objective

To check the existence of leakage and the effect of windowing on two close harmonic signals. Then to see when the two components can be resolved by the analysis (as introduced in Exercise 7.3), and show the importance of a logarithmic display in spectral analysis.

Reminder

Leakage exists when a signal is not periodic within the analysis window. Power existing in specific frequency regions is spilled (computationally) into neighboring regions, and can mask or modify the true power of components existing there.

Description

A signal composed of two harmonic components is generated. The following spectra are shown: the result of applying rectangular and Hanning windows (left and right plots), linear and logarithmic scales (upper and lower plots, Figure E7.4).

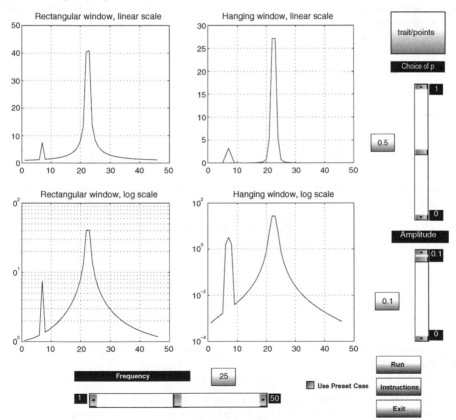

Figure E7.4

One component has a frequency coinciding with a computed one, hence no leakage, its amplitude controlled by the right hand lower vertical slider. A second component has a fixed amplitude, its frequency can be controlled via the lower horizontal vertical slider, but also by the parameter p (best by typing in the desired value, but also via the upper vertical slider). This parameter p determines the non-integer part of the number of periods for this component ($p = 10$ or $p = 1$ corresponding to an integer number of periods, $p = 0.5$ to an odd number of half periods etc).

Instructions/Tasks

Try to find the limit when the two components cannot be separated any more. First set p equal to zero (or 1), i.e. zero leakage. Change the amplitude and frequency of the first component, note the effect of the Hanning window.

Repeat for $p = 0.5$.

Exercise 7.5

Objective

To check the effect of zero padding.

Reminder

For
$$x = x(n) \quad 0 \le n \le N_1 - 1$$
$$= 0 \qquad N_1 \le n \le N - 1$$

i.e. appending $N - N_1$ zeros to the signal

$$X(k) = \sum_0^{N-1} x(n) \exp\left(-j\frac{2\pi}{N}ik\right) = \sum_0^{N_1-1} x(n) \exp\left(-j\frac{2\pi}{N}ik\right) + \sum_0^{N-N_1} 0$$

results in the DFT being unchanged. The computed frequencies and spacing are, however, changed from

$$\Delta f_1 = \frac{1}{N_1 \, \Delta t} \quad \text{to} \quad \Delta f = \frac{1}{N \, \Delta t}$$

with

$$N_1 \le N \quad \text{and} \quad \Delta f \le \Delta f_1 .$$

The uncertainty law: $\Delta f \, \Delta t \le C$ with C a constant

Description

Two types of signals can be chosen; the first is a transient half sine.

Changing the number of total points (by typing in the box) appends zeros to the transient. The upper plot in Figure E7.5 shows the transient (green) and appended zeros (blue). The lower plot shows the DFT, that of the original in green, that with appended zeros in blue. The second signal is composed of

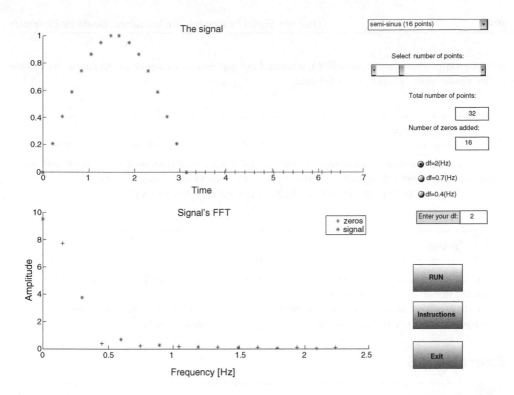

Figure E7.5

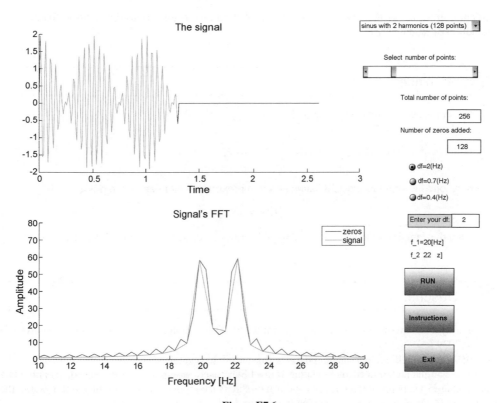

Figure E7.6

two harmonic components (Figure E7.6), separated in frequency by df, which can be chosen from three preset values, or by typing in a desired value.

Instructions

Choose the half sine transient. Compare the original DFT with results obtained with various numbers of appended zeros. Check the effect of the improved computational frequency resolution.
  Next choose the two component signals. Vary their separation frequency df.

Tasks

• For the half sine transient: check the effect of appending zeros on the computational frequency spacing of the spectral representation.
• For the two component signal: check the resolving power of the DFT needed to resolve the two components. Summarize the effectiveness of appending zeros for this task.

Exercise 7.6

Objective

To demonstrate the existence mechanism and control of bias errors for various signal types. To introduce the notion of random errors.

Reminder

The computational frequency spacing is

$$\Delta f = \frac{1}{N\,\Delta t}$$

For resonating systems, the 3 dB bandwidth is approximately (for damping ratios $\varsigma \ll 1$)

$$BW = 2\,\zeta f_0$$

with ζ the damping ratio and f_0 the undamped natural frequency (Appendix 4.A in Part B).

Description

A linear system, excited by white noise, is simulated. The sampling interval used in the analysis is $\Delta t = 0.0039$ [sec].
  An SDOF or 2DOF system can be chosen, with controllable eigenfrequencies and damping ratios. The PSD of the output is computed, by averaging M raw PSDs based on sections of N points. Some parameters can be chosen via pop-ups (frequencies and N), or sliders (dampings). However, the dampings and the parameter M can also be chosen by typing the required value in the appropriate box.

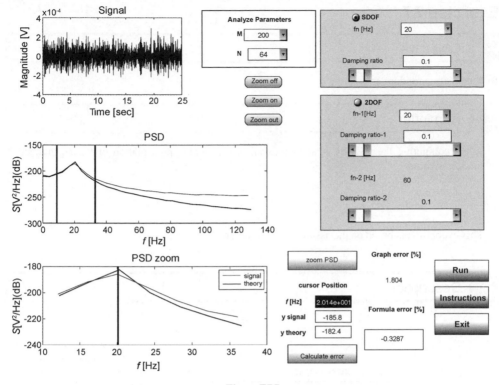

Figure E7.7

Three plots are shown in Figure E7.7, the output time signal (upper), PSD (middle) and zoomed PSD (lower). The zoomed range can be dictated by the two cursors on the middle plot. By choosing the option of zoomed PSD, a cursor can check the computed and theoretical values at any frequency. Then the computed and the theoretically estimated relative errors are presented.

Instructions

Choose an SDOF system, and $M = 200$ averages. Check the effect of N on the bias error at the resonance frequency. This should be repeated for various damping ratios and eigenfrequencies.

Repeat for the 2DOF system case.

Tasks

- For the SDOF system: find cases when large bias errors occur. If necessary choose specific damping values by typing the chosen values in the damping box. Show how such a task is affected by M.
- For the 2DOF case: investigate possibilities of having different bias errors for different spectral regions.

Exercise 7.7

Objective

To demonstrate random errors.

Reminder

When computing the PSD of a random signal, a practical approximation of the normalized random error is

$$e = \frac{1}{\sqrt{M}}$$

This expression is based on the assumption of random Gaussian signals.

Description

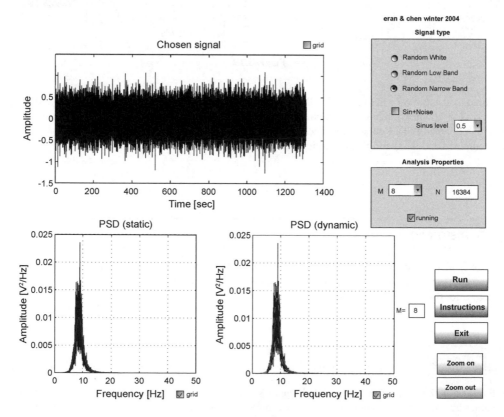

Figure E7.8

Various signal types can be chosen. The time signal is shown in the upper plot in Figure E7.8, the PSD in the lower left. Using the 'Running' option, a dynamic display of the PSD is shown in the lower right plot, with the PSD updated as additional signal blocks are made available. The number of blocks, M, can be chosen. After the analysis is performed, the theoretical and computed PSD can be seen in the lower right plot, together with the theoretical bands based on the predicted random error.

Instructions/Tasks

Choose a random white noise. For various M check (in the running mode) whether the random error for the computed PSD is predicted by the analytical formulae.

Repeat for a random low pass band signal. Check whether visual inspection could dictate when the result is reasonably acceptable.

Repeat for a noisy sine signal.

Exercise 7.8

Objective

To compute the PSD of a fixed duration signal, and show the resultant relation between random and bias errors.

Reminder

The frequency resolution via the FFT is

$$\Delta f = \frac{1}{N\,\Delta t}$$

For a fixed duration T_t, the number of averaged section is

$$M = \frac{T_t}{N\,\Delta t}$$

Hence $M/\Delta f = T_t$ is constant, and reducing the bias error (via Δf) will increase the random error (decrease the available M).

Description

Three types of signals are available: the response of an SDOF system to a random excitation, a noisy sine, and a noisy amplitude modulated carrier. The noise level is variable. The sampling interval used is $\Delta t = 0.007$ [sec].

The time signal is shown in the right plot, the PSDs in the left ones (Figure E7.9). The full range is shown in the upper plot. Choosing a specific band via cursors, the button 'Zoom PSD' will the show the zoomed range in the lower plot. Using the cursor of the lower plot, theoretical and computed PSD values are seen by the button 'Get y'.

Spectral analysis is performed by choosing N, thus affecting M, as the product $N*M$ is constant.

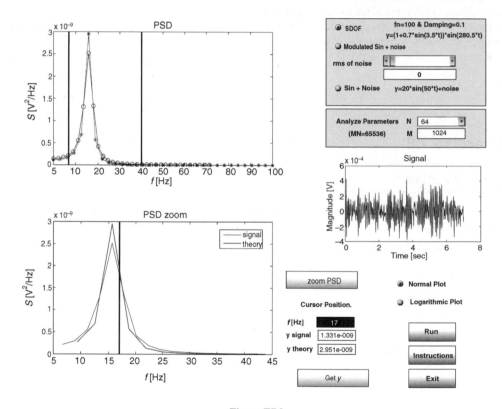

Figure E7.9

Instructions

- Case 1, the harmonic signal: its amplitude is 20 [V]. Set the noise level to 10 and use the log scale. Compute for $N = 64$ and 2048.
- Case 2a, the modulated signal: set the noise level to zero and use the linear scale. Compute for $N = 256$ and 512.
- Case 2b: repeat (2a) with the noise level equal to 8 and a log scale.
- Case 3, the SDOF case: set the noise level to zero, use the log scale. Compute for various values of N.

> **Tasks**

- Case 1: for the harmonic signal, determine the effect of changing N. Explain how the effect on the PSD is compatible with the results of Exercise 7.1b.
- Case 2: for the modulated signal, determine the minimum N required in order to identify sidebands (repeat Exercise 3.4 if necessary). As for case 1, explain the effect on the PSD values.
- Case 3: for the SDOF response signal, choose the analysis parameters in order to compute the PSD. It is necessary to use the zoom (and cursor) option in order to hone in on the desired range.

For the SDOF excited system, check that the results are compatible with those noted with Exercise 7.6. Specifically address the problem of the bias error. Summarize the results obtained for the spectral analysis when the signal duration is constant.

Exercise 7.9

> **Objective**

To investigate the relation between the PSD and autocorrelation of a random signal.

> **Reminder**

$$S(f) = F[R(\tau)] \qquad R(\tau) = F^{-1}[S(f)]$$
$$S(f) \leftarrow \rightarrow R(\tau)$$

> **Description**

The random signal is shown in the upper left of Figure E7.10. Two filtering options are available via buttons – LP (low pass) and BP (band pass). The frequency response of the filters is shown in the upper right plot, and is controlled via two cursors (the left cursor for the LP case, and two cursors for the BP case).

The lower plots show the filtered signal (left), its PSD (middle) and its autocorrelation function (right).

Note that after running the example, the cursors return to their original location. The analysis results (lower plots) are, however, for the cursor values chosen before.

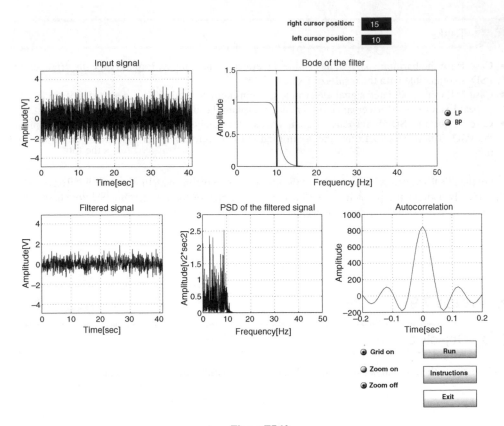

Figure E7.10

Instructions

Apply a low pass filter. Vary the critical frequency, observing the PSD and autocorrelation.
   Repeat for the band pass filter, varying the frequency region and bandwidth.
   Use the zooming option to investigate detailed patterns.

Task

Verify the law relating the PSD and autocorrelation function of a random signal.

Solutions and Summaries

Exercise 7.1(b)

For a sine wave, with a frequency coinciding with a discrete computational frequency ($k\Delta f$), the power in a single bin is constant, independent of N (Figure E7.11). It is necessarily equal to 0.5, the mean square value for a sine of amplitude 1. The PSD, however, being a density, equals the total power divided by Δf, and will vary with N (which affects Δf).

For a random signal, with a constant PSD within a chosen region, the power in a single bin will depend on Δf (hence on N), while the PSD is unaffected. Thus changing N will affect the PSD and the power in a bin differently for random or periodic signals.

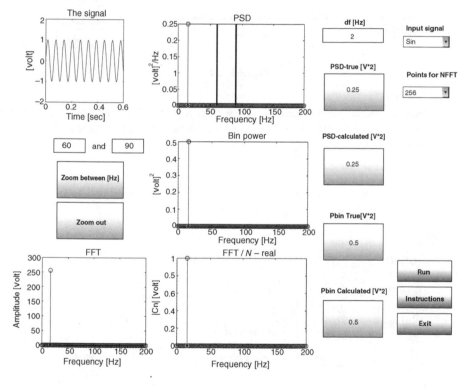

Figure E7.11

Exercise 7.2

For $K = 0$ or $K = 1$, an integer number of periods in the analysis window, there is no leakage. *For this case, the computational (analysis) frequency coincides with the true physical one.* The worst case occurs with $K = 0.5$, an odd integer numbers of half periods (i.e 3.5, 4.5, etc.). For a rectangular window (Figure E7.12) this results in two equal lines around the true frequency, with close to 40% underestimation of the peak, and a severe leakage.

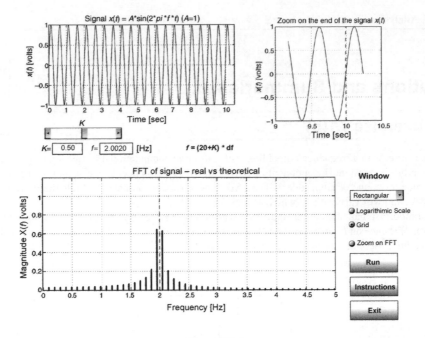

Figure E7.12

Using a Hanning window greatly decreases leakage (Figure E7.13). However, it broadens the result, i.e. results in some leakage for the case of $K = 0$, and a correction would be needed to get the correct magnitude at the signal's true frequency.

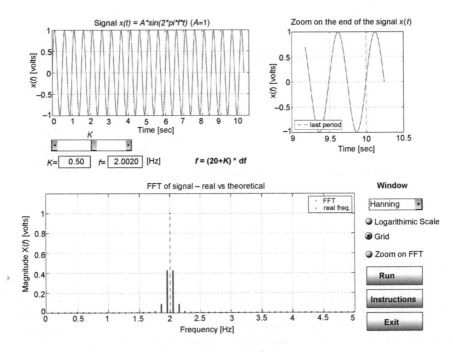

Figure E7.13

The flat top window has almost no scallop loss, and the magnitude is approximately constant as a function of K. This window would thus be insensitive to slight frequency fluctuations, where the true frequency shifts within the frequency analysis resolving bandwidth. The price paid is, however, more severe leakage.

To summarize the existence of leakage: this will exist whenever the signal does not span an integer number of periods. This is equivalent to the cases where the physical frequency does not coincide with computational analysis frequency. There are always two types of frequencies to consider – the *computational* ones, controllable by the analyst, and the *physical* one, dictated by the physical system.

Exercise 7.3

For $m = 2$, the two signal frequencies are such that their difference spans two computational frequencies on the FFT scale, and the two spectral lines will appear clearly.

Changing the spacing to $m = 1$ will result in the two lines coinciding with two adjacent FFT lines, and it will thus be impossible to identify them as being separate. Zooming in on the frequency scale (if necessary using the option of increased display resolution), shows this clearly. Only if the separation is at least $2\Delta f$ can two separated lines be recognized.

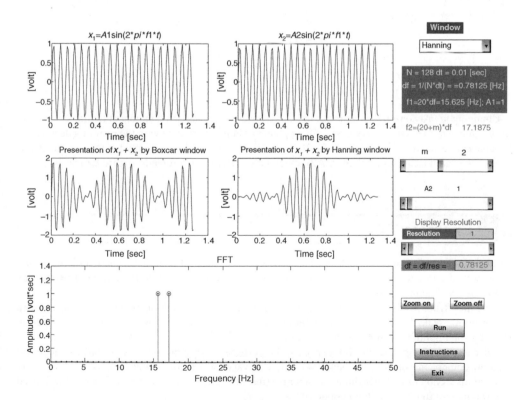

Figure E7.14

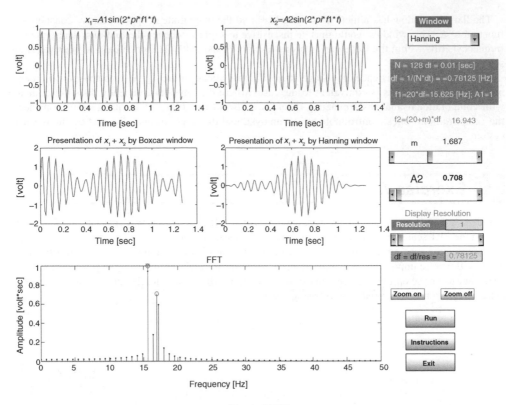

Figure E7.15

Changing to *m* a noninteger can result in leakage, causing the second component to hide completely the existence of the first one (try *m* = 2.5). Applying a Hanning window with the resulting decrease in leakage can reveal this component again. Both the effect of *m* (the leakage) and the ratio of the amplitude of the components (controlled by A2) affect the severity of the problem (Figure E7.15).

Exercise 7.4

As already shown by Exercise 7.3, leakage distorts the 'true' spectral pattern, and may limit the resolving power of spectral analysis. But a more severe effect can be noted from the choice of parameters shown below (Figure E7.16).

Here the existence of two components can only be seen when the Hanning window is used, and with a logarithmic scale. Using a window in order to reduce leakage results in an *increased dynamic range*, the range of amplitudes which can be discriminated.

For cases involving discrete (line) spectra, the purpose of spectral analysis could be to identify/ separate components of close frequency, possibly with varying order of magnitudes in amplitude. Only a visual inspection using logarithmic (or other possible nonlinear) scale could would reveal the desired spectral pattern (try also the preset option).

Another effect can be noted from the zero leakage case: the Hanning window broadens spectral peaks (see case in Figure E7.17), hence it would be detrimental for very close lines.

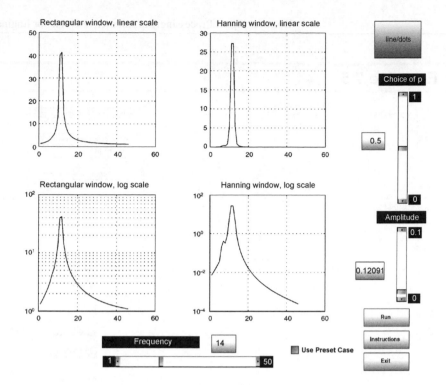

Figure E7.16

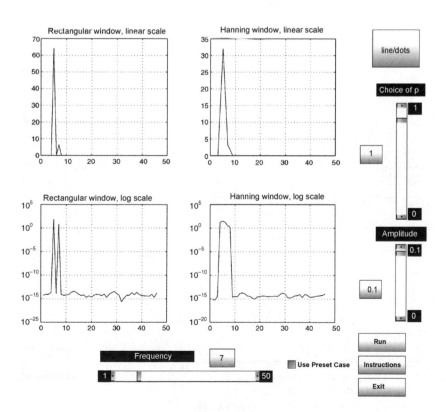

Figure E7.17

Exercise 7.5

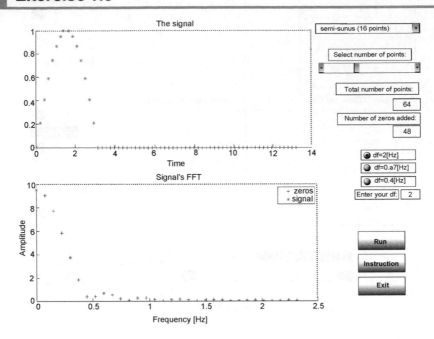

Figure E7.18

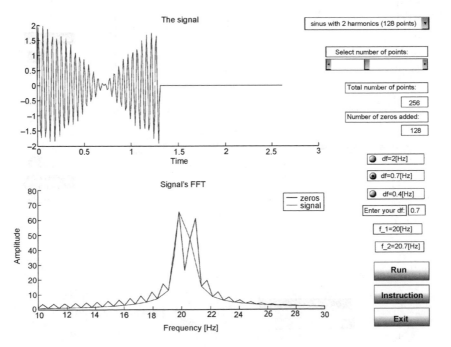

Figure E7.19

For the half sine signal, the effect of zero padding seems to be basically one of interpolation (Figure E7.18). Only a computational improvement in the frequency scale is obtained. Whatever the number of zeros appended, the original spectra will include the computed DFT samples.

Another demonstration of this effect is given by the two component case (Figure E7.19). According to the uncertainty law, a separation of 2 Hz in the frequencies will always be recognizable with a signal duration of approximately 1.4 sec, and never for a separation of 0.4 Hz, whatever the number of zeros appended. For a separation of 0.7 Hz, however, a marginal case, it is only through the appending of zeros that a fine enough frequency scale can enable the identification of two separate components.

Exercise 7.6

For an SDOF system, $fn = 20$, damping ratio 0.005, the bandwidth is 0.2. The computational frequency spacing for $N = 64$ is $1/(0.0039N) = 4$ Hz, resulting in a severe bias error. Using $N = 1024$ gives a much better estimation (Figures E7.20 and E7.21).

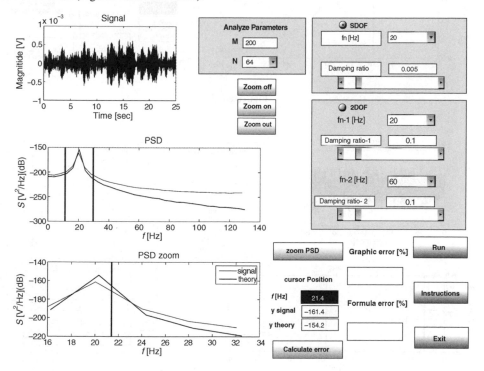

Figure E7.20

A much higher N is needed, to achieve a frequency spacing at least 1/3 of the bandwidth. For higher resonance frequencies and dampings, the 3 dB bandwidth is larger, resulting in a lower value for the minimal necessary N.

For a 2DOF system, the frequency spacing (dependent on the chosen N) is fixed for the whole range. The bandwidth of the two spectral peaks may differ however, and it is possible to have negligible bias errors in one region and significant ones in another. See for example Figure E7.22. Unless a high enough M is chosen, the random error may be so large as to make the PSD meaningless.

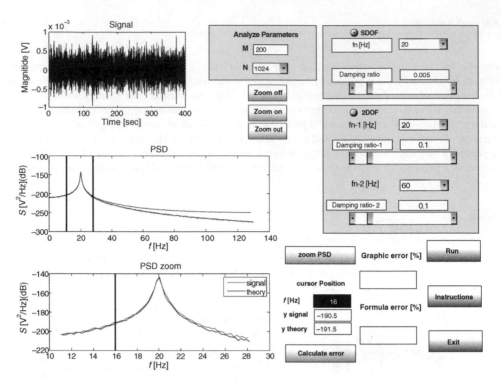

Figure E7.21

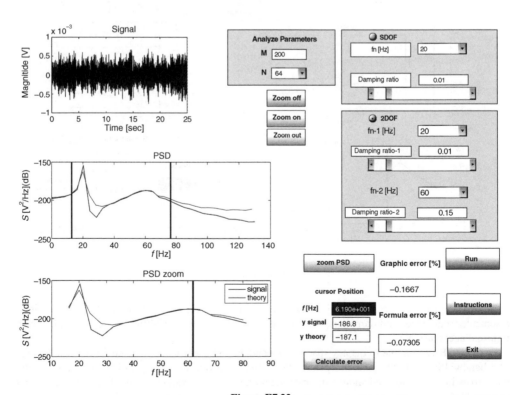

Figure E7.22

Exercise 7.7

As an example, running with $M = 8$, the expected normalized error is $1/\sqrt{8} = 0.35$. At the lowest frequency, we note a PSD (after eight averages) of 0.01, thus the expected bounds would be 0.0065–0.0135, in reasonable agreement with the result shown (Figure E7.23).

Using the running mode, visual inspection is very effective in showing the PSD shape in (almost) real time. This could be helpful when an a priori choice of M is difficult.

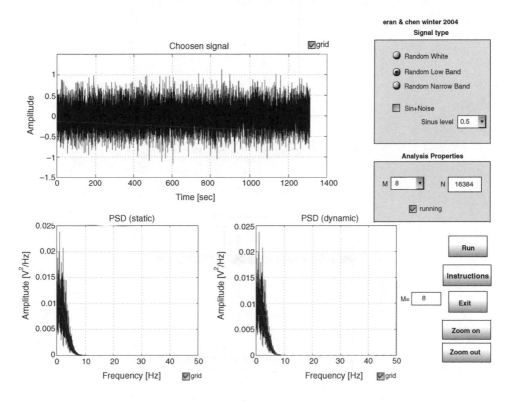

Figure E7.23

Exercise 7.8

• Case 1, the noisy sine signal (Figure E7.24): Changing N affects Δf. The PSD of the broad band noise is constant, independent of Δf; however, that of the harmonic signal is inversely proportional to Δf (see Exercise 7.1b). For example, with a sine amplitude of 20 (power equal 200), $N = 64$, we have $\Delta f = 1/(0.007*64)$, and the theoretical PSD is indeed $(20^2/2)/\Delta f = 89.6$, in perfect agreement with the result shown. For a noisy sine signal, in the bin containing the signal spectrum, the larger N, the better the signal /noise in the frequency domain.

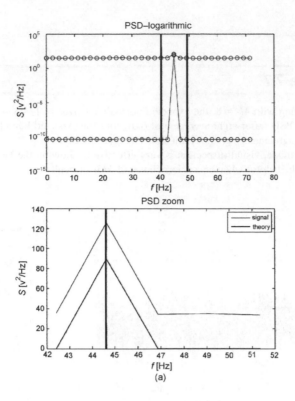

(a)

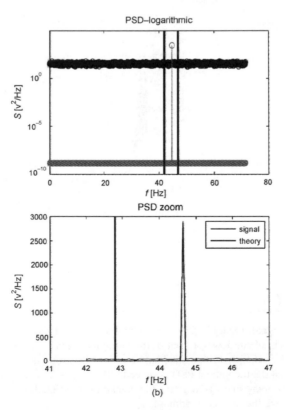

(b)

Figure E7.24

- Case 2, the modulated signal (Figure E7.25): The frequencies of the carrier and modulating are $f_c = 280.5/2\pi$ and $f_m = 3.5/2\pi$ respectively. For $N = 256$, the frequency resolution is $\Delta f = 1/256/0.007$, thus the spectral indexes for these two frequencies are $f_c/\Delta f = 80$ and $f_m/\Delta f = 1$, and no leakage is present. To separate the 80 Hz from the two sidebands, separated by 1 Hz, we must increase the resolution by a factor of at least 2 (note that a factor of 3 would have been recommended for any case involving leakage).

  With noise, the existence of the sidebands is masked, even with $N = 512$. However for noisy harmonic components, increasing N will increase the signal to noise ratio in the relevant spectral bin (as shown for case 1, the noisy sine signal). Increasing N for this modulated case (for example $N = 4096$ as shown), increase this signal to noise, and the sidebands can now be recognized (Figure E7.26b).

- Case 3, the SDOF system, excited by a random input (Figure E7.27): The 3 dB bandwidth is $2f_0\,\zeta = 2*15.9*0.1 = 3.18$ Hz. With $N = 512$, $\Delta f = 1/512/0.007 = 0.28$ Hz, and the bias error will be negligible. However with lower N, the bias error is noticeable, see for the case of $N = 64$, for which $\Delta f = 2.23$. M is dependent on N. For a random signal (the SDOF case with or without additive noise),

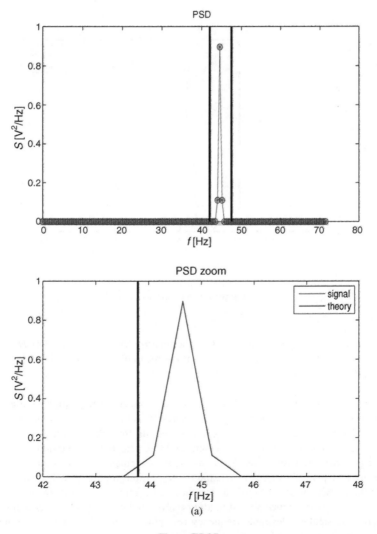

Figure E7.25

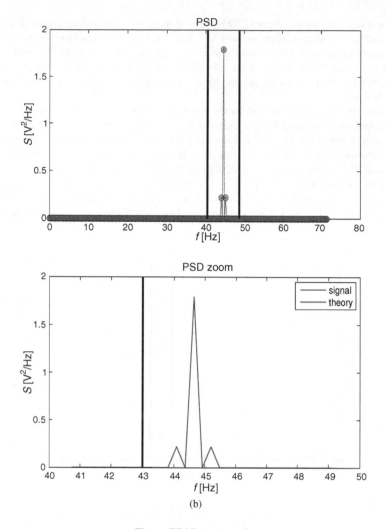

Figure E7.25 (*continued*)

a lower M (due to a higher N) will increase the random error. As an example, with $N = 4096$, a large random error exists. The bias error is very low, as the mean of the erratic noisy PSD seems to track the theoretical one (Figure E7.28).

A compromise between a high N (low bias and high random error) and a high M (high bias and low random error) may be indicated. A better option, if possible, is to use a longer signal duration.

To summarize: for a constant duration signal, computing the PSD via the segmentation method involves values of N and M which are not independent. Increasing the computational frequency resolution (increasing N) will decrease the number of averages of segment based spectra.

For a deterministic periodic signal, increasing the resolution (higher N) will increase the signal to noise ratio in the relevant spectral bins, and can thus be beneficial. For random signals, a balance is needed between bias error (higher N) and random errors (higher M). Both errors cannot be reduced simultaneously. The smallest desirable frequency resolution (hence the smallest N compatible with resolution requirements) is thus often advised.

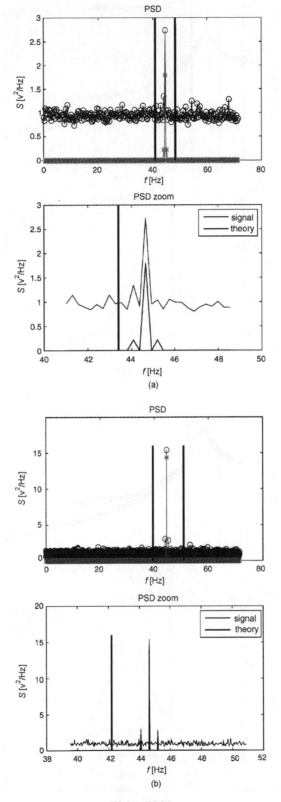

Figure E7.26

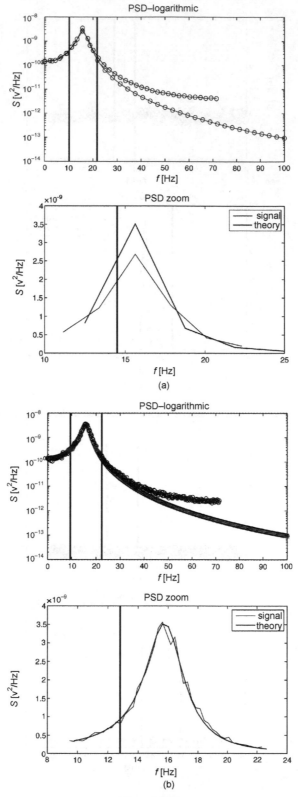

Figure E7.27

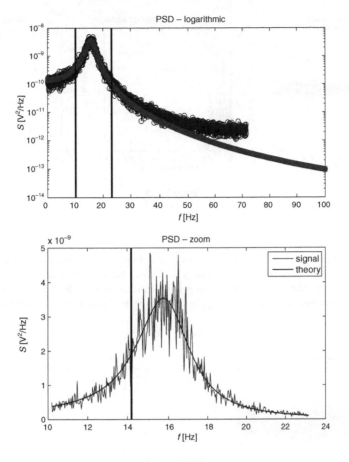

Figure E7.28

Exercise 7.9

For the LP case, approximating the filter by an ideal one, we have

$$S(f) = \begin{cases} S_0(f) & f \le f_0 \\ 0 & \text{otherwise} \end{cases}$$

$$R(\tau) = F^{-1}[S(f)] = S_0 f_0 \frac{\sin(2\pi f_0 \tau)}{2\pi f_0 \tau}$$

Choosing a critical frequency of $f_0 = 5$ Hz, we expect the first zero crossing of $R(\tau)$ to occur at 0.1 Hz, as indeed shown by Figure E7.29.

For the a narrow band BP case, we expect

$$S(f) = \begin{cases} S_0(f) & f_0 - \dfrac{BW}{2} \le f \le f_0 + \dfrac{BW}{2} \\ 0 & \text{otherwise} \end{cases}$$

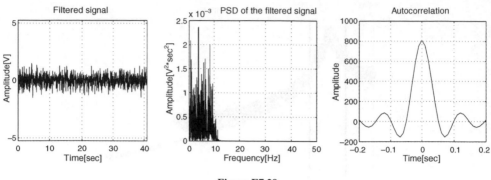

Figure E7.29

Choosing a band pass region of 20–30 Hz, we note (after zooming), the character of the filtered signal – an approximate constant zero crossing frequency of 25 Hz, the BP centre (see Figure E7.30). Such a constant zero crossing rate was already seen in previous exercises (2.1, the NB case). The autocorrelation function has an envelope approximating the $\sin(a)/a$ form, again with an oscillation of 25 Hz.

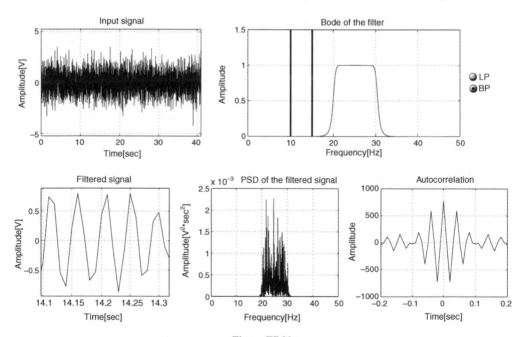

Figure E7.30

8

Envelope Detection

Overview

Only one exercise (8.1) is included. The technique is later applied in the diagnosis of roller bearings (Chapter 13).

The Exercises
Exercise 8.1

Objective

To compare the spectral analysis of signals and their envelopes for some specific signals.

Reminder

- For general modulated signals

$$x(t) = a(t) \sin (2 \pi f_c t)$$

- A narrow band random noise can be represented as

$$x(t) = a(t). \sin ((2 \pi f_c t + \phi (t))$$

with $a(t)$ a slowly varying random envelope, f_c the approximate zero crossing frequency and ϕ a random phase function.

Description

Signals can be chosen via the upper left pop-up menu. After choosing the one to be analyzed, 'Run' will load the signal (upper left in Figure E8.1) and its PSD (upper right). The two cursors in the spectral window can be used for filtering purposes, with the class of filter chosen by the 'Filter type' pop-up. The 'Execute' button will activate the filter. The lower left plot then shows the filtered signal and its envelope

Discover Signal Processing: An Interactive Guide for Engineers   S. Braun
© 2008 John Wiley & Sons, Ltd

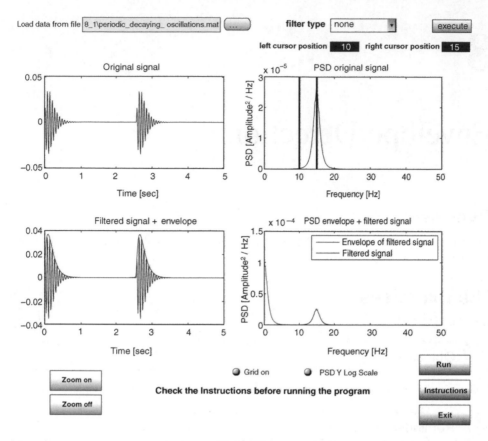

Figure E8.1

(in red), while the lower right plot shows the PSD of the filtered signal as well as that of the envelope (in red). A log display is also available when an improved visual inspection of spectra is necessary.

Instructions

Choose the periodic decaying oscillation signal. Using 'None' filtering option, run the program. Repeat for modulated_signal_1. Repeat for modulated_signal_2. Next repeat applying a band pass filter rejecting the two highest frequency components.

Choose the random signal. Using 'None' filtering option, run the program. Next apply a narrow band filter.

Tasks

For the periodic decaying oscillations, the modulated_signal_1, the raw and filtered modulated_signal_2, describe and interpret the signals, envelopes and their spectra. It may be useful to run Exercise 3.4 again in order to recall properties of amplitude modulated signals.

Repeat for the raw and filtered random signal. Run again Exercise 2.1, choosing the random NB signal, and summarize the characteristics of such signals.

Solutions and Summaries

   Exercise 8.1

For the periodic decaying excitation, the signal spectra are narrow band, centred around the frequency of the transient. The envelope is a low frequency signal, as is also evident in the spectra (Figure E8.2).

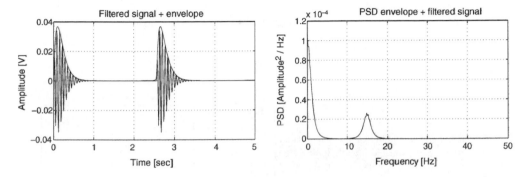

Figure E8.2

For the modulated signal_1, there is a modulation frequency of 1 Hz with a carrier frequency of 8 Hz, with components of 8 ± 1 Hz. The envelope frequency is obviously 1 Hz (Figure E8.3), and a similar result is obtained for the modulated_signal_2, however with higher carrier and modulating frequencies (Figure E8.4). There are now fewer carrier periods per modulation period, and the envelope detection is less ideal. The envelope spectra show not only the fundamental frequency of the envelope, but (using the log scale) also its harmonics.

Rejecting the carrier and higher sideband, via a suitable filtering operation, gives the results shown in Figure E8.5. This is actually close to a simple harmonic signal, with a DC envelope.

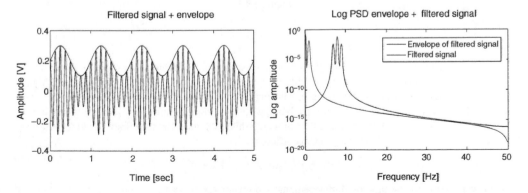

Figure E8.3

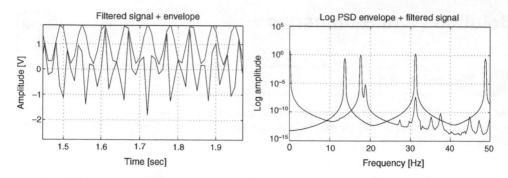

Figure E8.4

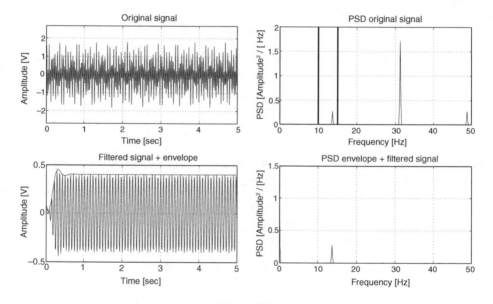

Figure E8.5

For the random signal we get Figure E8.6. Both the signal and its envelope are random, the envelope does not track the signal exactly, only roughly its extrema, hence its spectrum is attenuated at higher frequencies. Figure E8.7 shows the result of filtering in order to get narrow band characteristics. The spectra show that the signal is indeed narrow band, with zero crossing frequency roughly in the middle of the band pass region (zooming in will help in seeing this) as dictated by the cursors used for filtering. The envelope is low pass, slowly varying, according to

$$x(t) = a(t) . \sin(2\,\pi . f_c t + \phi\,(t))$$

This is typical of narrow band random signal, as also seen in Exercise 2.1.

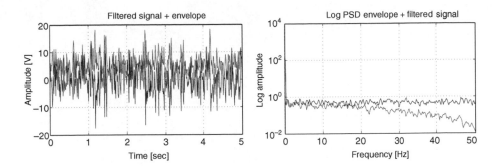

Figure E8.6

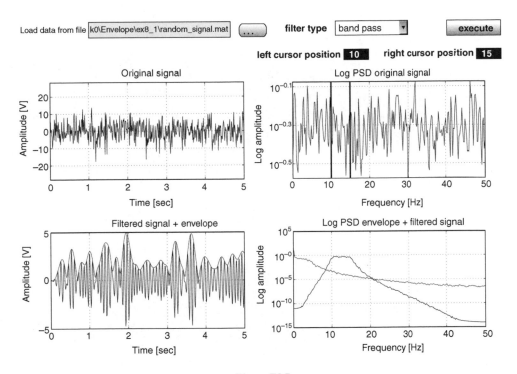

Figure E8.7

9

The Spectrogram

Overview

One of the simplest combined time–frequency analyses, the spectrogram, is addressed by two exercises. The first, 9.1, introduces the problematic time versus frequency resolution aspect. In Exercise 9.2 two classic test signals (single and dual chirps) can be analyzed, in addition to an experimentally acquired signal for which the spectrogram's limitations are even more evident.

The Exercises
Exercise 9.1

Objective

To introduce a time–frequency representation based on the spectrogram. To understand the trade-off between resolution in the time and frequency domain.

Reminder

- The time–frequency description is based on computing an FT of a (sliding) windowed signal:

$$S_x(t,f) = \int_{-\infty}^{\infty} x(u).h(u-t).e^{-2j\pi ft} du$$

- The resolution in time and frequency is dictated by the window function

Description

Various signals can be analyzed, external, two built in and any added to the path. The sampling intervals are 10 and 0.02 msec for dat9_1 and dat9_2 respectively. The signal is seen in the lower left plot in Figure E9.1, and the spectrogram in the upper left one. nfft, the number of points in each FFT, is controllable. Positioning the cursor in the spectrogram defines a time slice of nfft points centred around it, the spectrum of which is shown in the upper right plot by applying the 'Update spectrum'

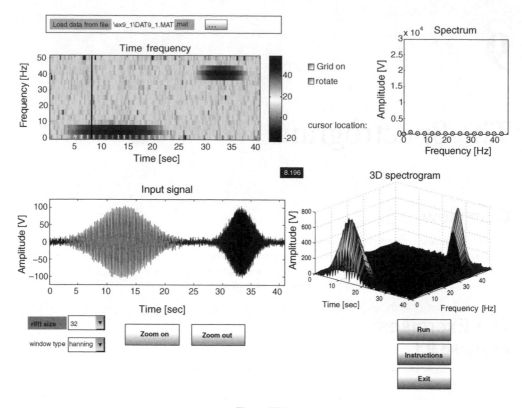

Figure E9.1

button. This spectrum approximates an *instantaneous spectrum*. The cursor's location is displayed in the appropriate box.

A 3D time–frequency plot is shown in the lower right plot. The perspective can be controlled via the 'Rotate' option.

Instructions/Tasks

Choose the signal dat9_1, nfft = 128 and run program. Using the cursor, check the spectrum around time locations of 14 and 32 sec.

Inspect the spectrogram, and repeat for various nfft. Complement the inspection by viewing the 3D plot from different perspectives, including ones that coincide with the spectrogram.

The objective is to see the aspects of resolution in the time and frequency domain. Repeat for dat9_2. Check again exercise 5.2, where filters could be used in order to analyze the same data (dat5_2).

Exercise 9.2

Objective

To synthesize approaches in time, frequency and time/frequency domains for some specific signals.

Description

The program performs a multitude of analyses. The raw signal is depicted in Figure E9.2, together with its envelope (in red). Also shown is the spectrum (lower left) and the spectrogram (lower right). The FFT block size as well as the overlap can be controlled. A cursor in conjunction with zooming (see 'Help') enables us to hone in on signal and envelope details. The ordinates at the cursor location appear in the box. For the time/frequency analysis, 2D or 3D plot are available, the 3D can also be rotated.

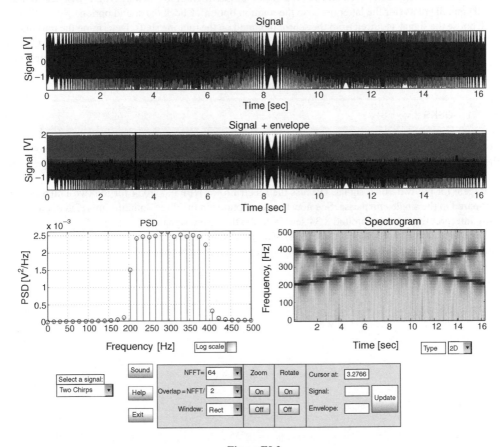

Figure E9.2

Three types of signals as well as an external one can be analyzed. These consist of (1) a chirp signal whose frequency increases linearly with time, (2) a combination of two linear chirps with increasing and decreasing frequencies respectively, and (3) an experimentally acquired signal, entitled 'Cardan', from a rotating machine. This is the output of an electromagnetic encoder, generating signal pulses at the rate of fixed periods per shaft rotation. Due to a faulty alignment, the shaft speed is a function of its radial location, thus varying periodically as well (with a period equal to that of the shaft). Hence the frequency as well as the amplitude (in an electromagnetic generator, output is proportional to velocity) of the generated signal varies periodically. Using 'Sound' it is also possible to listen to an aural representation of the signal, using the computer's sound card.

Instructions

1. Choose single chirp signal. Check the spectrum for NFFT = 64 and 1024. For these values, check the time and frequency resolution of the spectrogram. Inspect both the 2D and 3D plots (for the latter use also the rotation option). Check the effect of overlap on the time resolution. Repeat for NFFT = 256. Apply a rectangular and Gaussian window. Check the sound option.
2. Choose the two-chirp signal. Using the zoom option, check the signal and envelope structure. Check the spectrum for NFFT = 64 and 1024. Check the spectrogram for various NFFT. Inspect both the 2D and 3D plots (for the latter use also the rotation option). Check the sound option.
3. Choose the signal acquired from the sensor mounted on the rotating shaft. Determine the maximum, minimum and averaged frequency of the signal generated, as well as that of the envelope. Choose NFFT in the range of 64–2048 and see the effect on the spectrum.. Repeat for the spectrogram (2D and 3D representations). Check the sound option.

Tasks

For the single chirp, explain the spectra for various NFFT. Explain the time and frequency resolution of the spectrogram, and summarize the effect of switching between a rectangular and Gaussian window. See how the descriptions in the three domains as well as the sound option are related.

Repeat for the two-chirp signal. Discuss the spectrum for NFFT > 1024 for the two-chirp case as compared to the single chirp case. See how the descriptions in the three domains are related, and carefully interpret the behavior around 8.34 sec. Describe the sound signal.

For the 'Cardan' case, inspect the time signal and envelope patterns and predict expected spectral components. Explain the spectrum, checking for predicted components.

Solutions and Summaries

Exercise 9.1

Using the cursor, the relevant spectrum shows regions of 4 and 40 Hz. The time domain shows a span of approximately 5–20 sec for the first component, and 38–48 sec for the second. This can be noted both from the spectrogram and the original time plot. Reducing nfft increases the temporal resolution and smears the frequency range of the spectrogram.. The 3D spectrogram display can be rotated to show the projection on the frequency plane, showing mainly two spectral peaks (Figure E9.3).

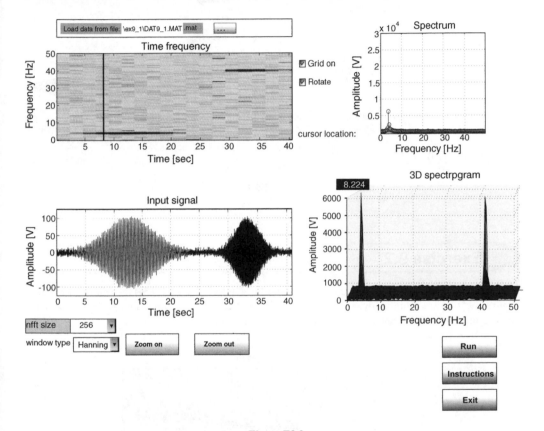

Figure E9.3

Using nfft = 128 it easy to see the effect of using specific windows. The boxcar one shows leakage, a spread in the frequency domain. The Gaussian one has a good compromise for both domains, and the decrease in leakage is evident.

Analyzing dat9_2 , the time frequency structure of the signal is easily understood from the 2D and 3D representations (after suitable rotation) of the spectrogram (Figure E9.4). The low frequency component exists in the complete time interval analysed, concentrated around 5000 Hz. The sharp transient around 6 msec has energy in a high frequency band around 42 kHz. Comparing this to Exercise 5.2, we would have needed to perform two filtering operations around these two frequency regions (determined by a spectral analysis) in order to understand the time pattern. The Gaussian window gives a higher dynamic range (background color in the upper right area is bluer) but a widening of the main lobe.

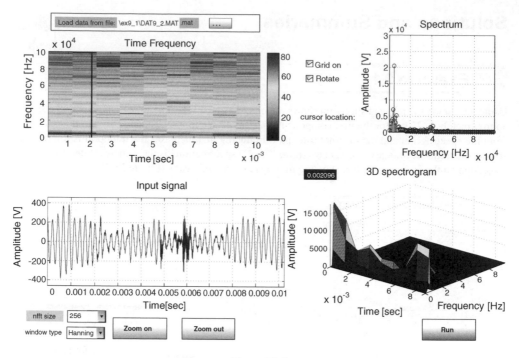

Figure E9.4

Exercise 9.2

For the single chirp signal: compared to the rectangular window (Figure E9.5a) the Gaussian window (Figure E9.5b) gives a higher dynamic range (background color is bluer) but a widening of the main lobe (Figure E9.5). This may be easier to notice for a small NFFT.

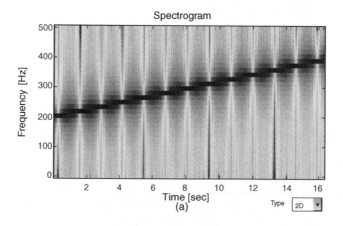

Figure E9.5

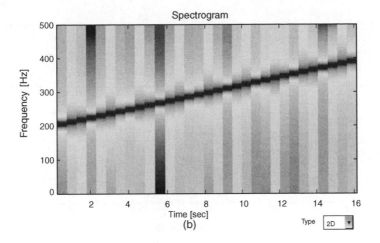

Figure E9.5 (*continued*)

The 2D and3D views show a single component, clearly increasing in frequency and of constant magnitude. In the time domain, zooming in, we note shorter periods with increasing time. The spectrum just shows the band over which the energy is distributed, certainly not any time trend. The audible sound demonstrates an increasing pitch.

Increasing NFFT gives a finer frequency resolution, time resolution depends on NFFT and overlap. For NFFT > 1024, the spectrum shows a number of separated bands. This is due to the way the spectrum is computed, via segment averaging. For successive segments, the frequency range increases, and for the specific segments of this case, each segment covers a nonoverlapping range. The final spectrum is computed by averaging segment spectra, hence the appearance of separated spectral bands (Figure E9.6). The audible sound clearly demonstrates the increasing frequency.

Similar results are noted for the combination of two chirps, one ascending, the other descending. two chirps The result around 8.34 sec, however, needs a more critical interpretation. Only

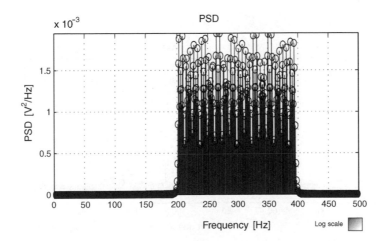

Figure E9.6

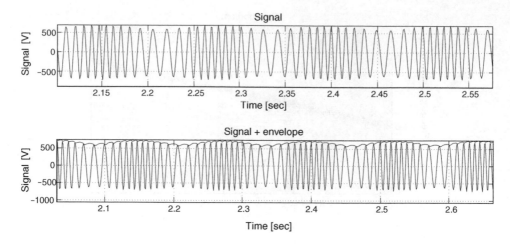

Figure E9.7

one specific frequency can be defined at the crossover between the two time/frequency lines. The time signal shows a beating, easily noticeable as the frequencies get closer, and the difference between them changes sign. A peak in the 3D presentation at this time indicate a summing up at a single frequency. Audibly this point is also recognizable. The spectral plot for NFFT > 1024 again shows the separate bands, but also that the temporal behavior is certainly completely invisible! The audible sound is again complex, but with repeated listening, the time frequency pattern can be recognized.

For the 'Cardan' signal, the time signal shows both amplitude and frequency modulations (Figure E9.7). The envelope has a period of 0.125 sec, hence a modulation of 8 Hz.

From the spectrum, a 8 Hz spacing between lines is indeed seen (Figure E9.8). Multiple side-bands, and higher sidebands of greater amplitude than lower ones, are typical for FM, as was en-countered in Exercise 3.4. They are not symmetric, supporting the fact that both amplitude and frequency modulation exist. The carrier frequency is more difficult to determine from the time signal; however, the mean period (using the longest and shortest) is approximately 10 msec, hence a carrier of 100 Hz.

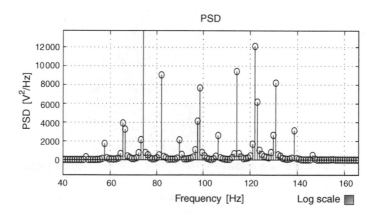

Figure E9.8

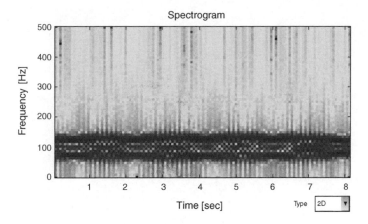

Figure E9.9

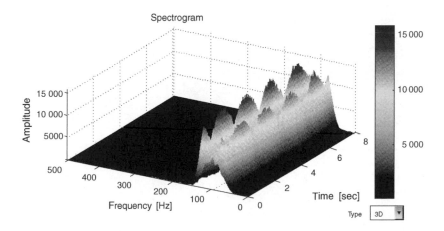

Figure E9.10

The spectrogram will show the sidebands for NFFT equal to say 512. For shorter NFFT, with insufficient frequency resolution, the sideband pattern disappears, but on the other hand the amplitude modulations, before unnoticeable are now apparent. Only an intermediate NFFT = 128, as seen in Figure E9.9, could (for these specific data) show both. One good way to inspect this is by a rotation of the 3D spectrogram, using a Gaussian window, which has the best time/frequency compromise (Figure E9.10). The audible sound reveals mainly the amplitude modulation.

10

Sampling

Overview

Signals first have to be acquired before they can be processed. Some errors, like aliasing, cannot be controlled after being introduced within the sampling process. Errors introduced by the sampling process are demonstrated in this chapter. Exercise 10.1 introduces quantization errors, and the need to match dynamic ranges of the instrumentation to that of the signal is addressed by Exercise 10.2. The aliasing effect is introduced by Exercise 10.3, and its control via an anti-aliasing filter by Exercise 10.4.

The Exercises

Exercise 10.1

Objectives

To demonstrate the quantization error.

Reminder

For an N bit bipolar ADC, the quantization step is

$$\Delta V = \frac{V_{FS}}{2^{N-1}}$$

Description

Shown in Figure E10.1 is a sampled harmonic signal. The actual sampled values are shown by red circles. Two signal classes, of large and small amplitudes, can be chosen via a pop-up. The specific signal amplitudes (A), sampling frequencies and the number of bits used for digitizing can also be chosen for each of the two classes. An error signal is computed as the difference between the true signal values and the sampled value, the output of an N bit digitizer, and shown superimposed on the sampled signal plot. The maximum and RMS value of the error signal are noted below the plot.

Discover Signal Processing: An Interactive Guide for Engineers   S. Braun
© 2008 John Wiley & Sons, Ltd

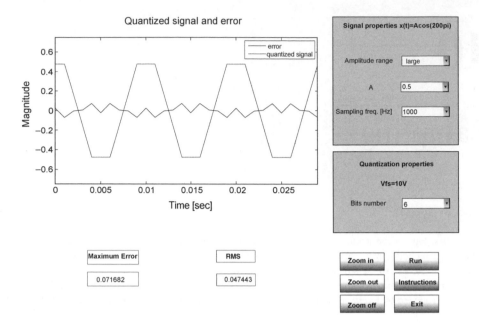

Figure E10.1

<table>
<tr><td style="background:gray;"> </td><td>

Instructions

</td></tr>
</table>

Run the program with the following parameters:

- Large signal:

$A = 0.5$	Nbits $= 6$	$f_s = 500$
$A = 0.5$	Nbits $= 6$	$f_s = 2000$
$A = 0.5$	Nbits $= 12$	$f_s = 500$
$A = 0.5$	Nbits $= 12$	$f_s = 2000$

- Small signal:

$A = 0.005$	Nbits $= 12$	$f_s = 500$
$A = 0.005$	Nbits $= 16$	$f_s = 2000$

<table>
<tr><td style="background:gray;"> </td><td>

Tasks

</td></tr>
</table>

Summarize the results. Next define a desired approximate accuracy, say 2%. For a digitizer, what would then be the minimum number of bits? What is a reasonable sampling frequency for the given signal?

Exercise 10.2

<table>
<tr><td style="background:gray;"> </td><td>

Objective

</td></tr>
</table>

To show the effect of matching the dynamic ranges of the data acquisition system to that of the measured signal.

Reminder

The quantization error is minimized by matching the signal level applied to that of the ADC.

Description

A transient is acquired with an acceleration sensor, with a sensitivity of 1[V/G]. Two measurements are undertaken, with different setting the data acquisition. Some specifications of the data acquisition systems appear in the upper right box (Figure E10.2). A preamplifier, preceding the digitizer, is part of the data acquisition system. Two settings – high and low gain – are available.

The output using a low gain setting is shown in the two upper plots, that with the high gain in the lower plots. The plots are those of the digital data, after applying a calibration factor depending on sensor sensitivity and the gain of the data acquisition system, thus showing approximately the same transient levels. The quantization level can be checked in the right plots by cursors, with the temporal location of the chosen points displayed as x1 and x2. Specific signal times can be accurately chosen by using the zoom options. By operating 'Run', the chosen level differences of the digitizer's output are displayed as dl (low gain amplifier setting) and dh (high gain amplifier setting) as well as their ratios. Note that the values shown are the digitizer's actual output (in volts), not that of the computed acceleration which is shown on the plots. An automatic option will choose preset signal locations.

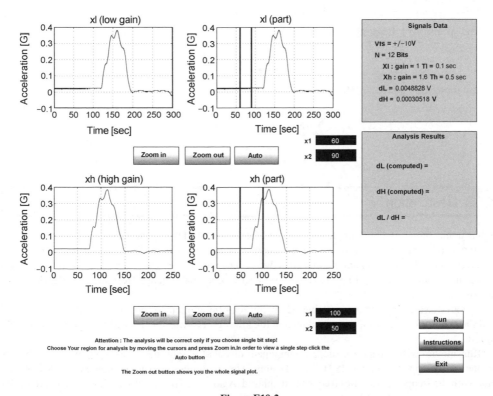

Figure E10.2

> ## Instructions
>
> Zoom in around the signal maxima (around 150 sec for the upper plot and 160 sec for the lower one). By carefully zooming in and moving cursors repeatedly, locate the two cursors such that single quantizations step are chosen. Using 'Run' compute the ratio of the two steps.

> ## Tasks
>
> Find the relation between the amplifier's gain ratio to that of the quantization steps. Check the results by switching to the automatic mode.

Exercise 10.3

> ## Objective
>
> To demonstrate the aliasing effect for signals with multiple harmonic components. To check the effectiveness of using antialiasing filters.

> ## Reminder
>
> The spectrum of the sampled signal is
>
> $$X_s = \frac{1}{\Delta t} \sum_{k=-\infty}^{k=\infty} X\left(f - \frac{k}{\Delta t}\right)$$
>
> and for a sine signal of 10 Hz
>
> $$\frac{1}{\Delta t} \sum_{n=-\infty}^{n=\infty} \left[\delta(f - 10 - nf_s) + \delta(f + 10 + nf_s) \right]$$

> ## Description
>
> A harmonic signal of fixed amplitude (10) and frequency (1 Hz) is generated. Two harmonic components can be added to it, with controlled amplitudes and frequencies via the pop-ups in the signal generation box. The total signal is sampled at 50 Hz. The composite signal and its spectrum are shown in the upper plots (Figure E10.3). Color of the spectral lines will change from blue to red when they are aliased.
>
> An antialiasing filter of controllable bandwidth can be applied to the signal. The filtered signal and the spectrum after sampling are shown in the lower plots. Aliased spectral lines are again shown in red.
>
> 'Run movie' will generate a second component of equal amplitude to the first fixed one, but its frequency is stepped from 2 to 48 Hz in 2 Hz steps. The changing spectrum is shown in the lower right plot, with the stepped (dynamic) frequency displayed. Again, the spectral line will switch to a red color once aliasing occurs. The filter does not operate in the movie mode.

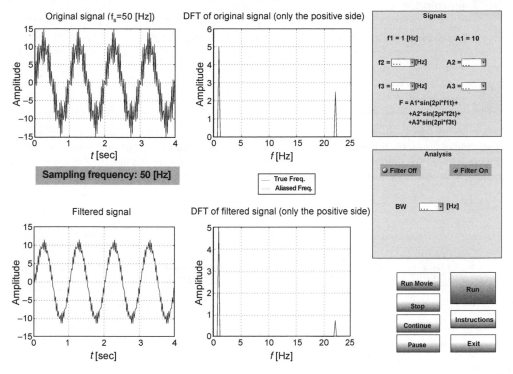

Figure E10.3

Instructions

Set filter to off. Set $A_2 = 20$, $A_3 = 0$, step f_2 from 4 to 36 (in steps of 4), each time activating 'Run'. With filter on, $A_2 = 10$, $A_3 = 20$, $f_2 = 18$ and $f_3 = 32$, see effect of changing the filter's BW from 4 to 45. Next run 'Movie'.

Task

- Summarize the aliasing effect.
- Check the rule for aliased frequencies when $f > f_s/2$.
- Explain the effect of a real antialiasing filter, as opposed to an ideal one.

Exercise 10.4

Objective

To investigate how the antialiasing filter properties affect the sampled signal.

Reminder

The antialiasing filter modifies the signal prior to the sampling operation.

Description

Two types of signals can be chosen. The first is a low frequency (1 Hz) square signal, the second consisting of the sum of this square signal and a sine of higher frequency (33 Hz) and smaller amplitude (Figure E10.4). The sampling frequency can be controlled via a pop-up. An antialiasing filter can be applied, with critical frequencies of 10 or 12 Hz for the first and second case respectively.

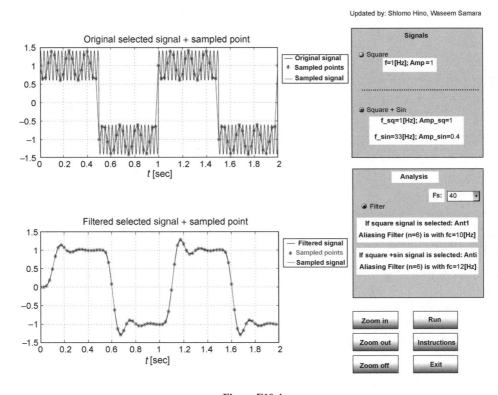

Updated by: Shlomo Hino, Waseem Samara

Figure E10.4

Instructions

(a) Choose the square + sine signal. Run the exercise with the antialiasing filter off, for sampling frequencies of 10, 20, 40 and 40 Hz.
(b) Repeat for the square signal.
(c) Repeat for the square signal with the antialiasing filter on.
(d) Repeat for the square + sine signal with the antialiasing filter on.

> ## Tasks

- Explain the results for cases (a) and (b).
- Explain the result for case (c).
- Explain the result for case (d).
- Summarize the effect of realizable antialiasing filters, discussing the necessary filter parameters.

Solutions and Summaries

Exercise 10.1

For the large signal, with $A = 0.5$, using Nbits $= 6$ results in a large quantization error (see Figure E10.5). The maximum error is close to 0.07, which is approximately 15% (0.07/0.5). Using 12 bits reduces the error to approximately 1.5%. Computing the quantization step for $N = 6$ results in $10/2^{11} = 0.0049$, providing a good estimate for the error. However, an appropriate sampling frequency seems required, beyond 500 Hz, and $f_s = 2000$ Hz seems sufficient.

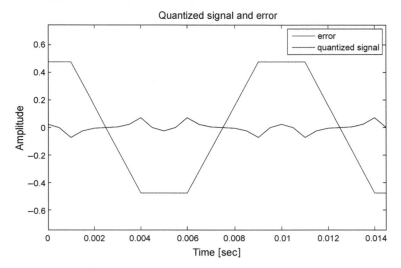

Figure E10.5

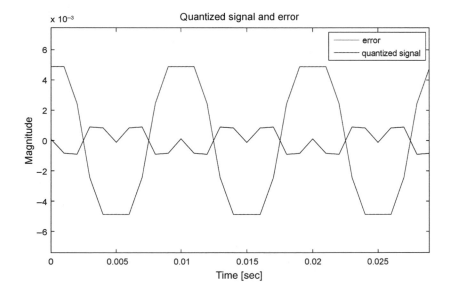

Figure E10.6

For a small signal, a higher number of bits are required, and the 12 bits of the first four cases is not sufficient (Figure E10.6). Using 16 bits gives a quantization step of $10/2^{15} = 3.0518\mathrm{e}{-}004$, the estimated error being $0.00030518/0.05 = 0.006$. Again, a sampling frequency of 200 Hz seems indicated. (A quantitative approach to the correct sampling frequency will be tested in Exercises 10.3 and 10.4.)

Exercise 10.2

By zooming in on the recommended time zones, we get results as shown in Figure E10.7 and the quantization steps can be seen clearly. The temporal location, chosen by the cursors, appear in the blue boxes. A single quantization step is now chosen (Figure E10.8). Applying 'Run' prints the quantization steps (left frame), which equals those in the digitizer's specification (upper right frame). Also printed is their ratio (16), and hence the amplifier gain equals 16.

The use of a high gain results in a smaller quantization step. The correct choice is thus to match the range of the digitizer to that of the signal! *A too low analog input to the digitizer can result in so-called quantization noise.*

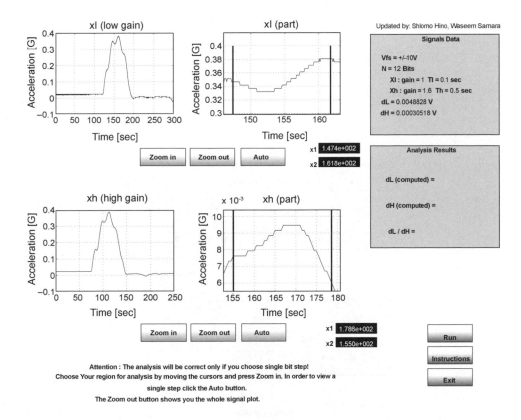

Figure E10.7

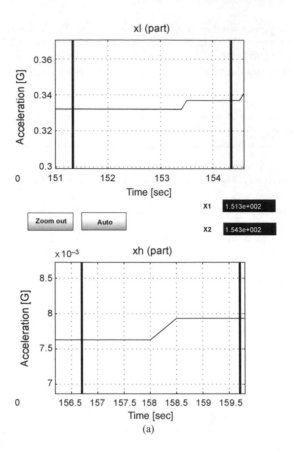

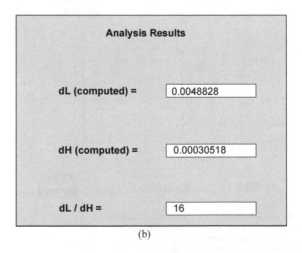

Figure E10.8

Exercise 10.3

With $A_2 = 20$, $A_3 = 0$, $f_2 = 28$ and the filter set at 25, we get results as shown in Figure E10.9. Aliasing occurs for components of frequency $f > 25$ Hz, half the sampling frequency. A 28 Hz component will appears at 22 Hz, and the nonideal filter, set to 25 Hz, will attenuate it only to 30%.

For frequencies of $f = 28$, 30, 32 and 34 Hz, we get aliasing to frequencies of $f_{alias} = 22$, 20, 18 and 16 Hz respectively, corresponding to $-f+2*f_s$. This is in accordance with Equation (10.2) in part B (with $k = 2$), remembering that the original signal, in the two-sided representation, has frequencies of $\pm f$.

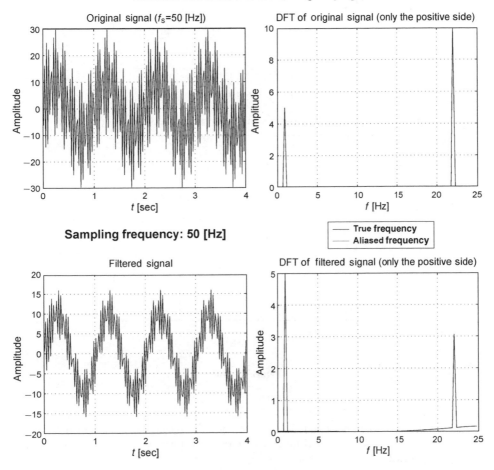

Figure E10.9

For $A_2 = 10$, $A_3 = 20$, $f_2 = 18$ and $f_3 = 32$ we get an aliased frequency coinciding with the second frequency. The nonideal filter, set at 25 Hz, does not attenuate it completely. The dynamic display achieved by the 'Movie' mode shows the onset of the aliasing effect (Figure E10.10). It occurs when

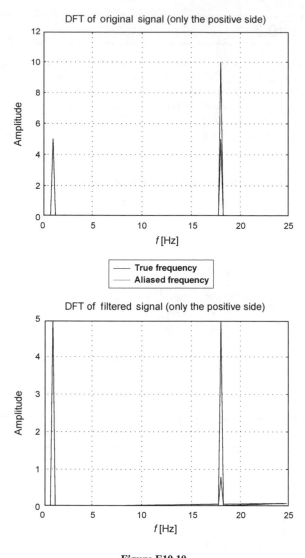

Figure E10.10

whenever $f > f_s/2$ (25 Hz in our case). The antialiasing filter should be set theoretically to a critical frequency of $f_s/2$. As only nonideal filters can be realized (Chapter 5), a slightly smaller critical frequency is indicated.

Exercise 10.4

Using $f_s = 20$ gives results shown in Figure E10.11 and the sine signal of 33 Hz is aliased, as noticed by the lower frequency of the signal superimposed on the square signal. The estimated aliased frequency

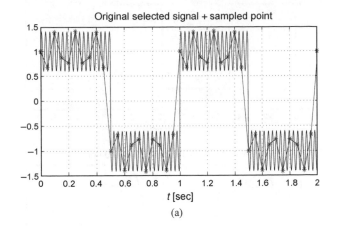

Original selected signal + sampled point

(a)

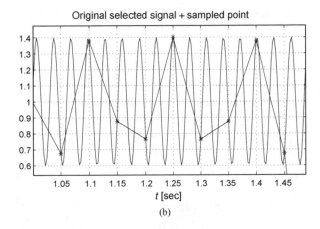

Original selected signal + sampled point

(b)

Figure E10.11

(see Exercise 10.3) should be $-33+2*f_s = 7$ Hz. The corresponding period is 14 msec, in reasonable agreement with this prediction – see the zoomed-in plot (Figure E10.11b). To avoid aliasing, we would need at least $f_s > 66$ Hz.

For both the square and the square + sine signal, sampled at 40 Hz, we get results shown in Figure E10.12. The sampling of the square signal seems adequate. However, when the antialiasing filter is applied, the result is a modified shape. As this is due to the filter only, we conclude that this modification is due to the dynamic response of the filter at the discontinuities of the square wave. A similar result occurs for the square + sine signal. The high frequency of 33 Hz is rejected, and aliasing avoided. However, the dynamic response of the filter still affects the result.

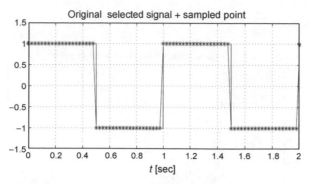

Original selected signal + sampled point

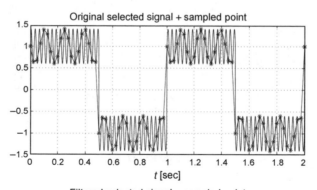

Filtered selected signal + sampled point

(a)

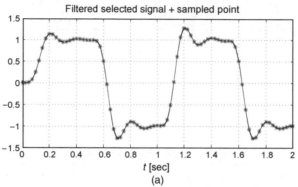

Original selected signal + sampled point

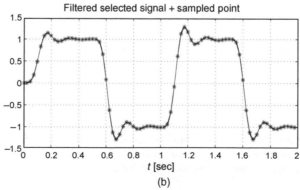

Filtered selected signal + sampled point

(b)

Figure E10.12

The specifications of the filter include the filter type, the critical frequency and the steepness (often specified via the filter's order), as encountered in Chapter 5 and the relevant exercises of that chapter. The filter will affect the response to transients. The steeper the frequency response, the longer the impulse response. Sometimes a linear phase filter may be required.

11

Identification – Transfer Functions

Overview

All exercises in this chapter investigate the case of linear systems excited by random inputs, with the transfer function computed via FFT based methods. The main intent is understanding and controlling errors involved in the identification process.

Exercise 11.1 deals with the effect of insufficient resolution, and Exercise 11.2 with the effect of delays between input and output. Exercises 11.3 and 11.4 deal with the effect of additive measurement noise as well as the choice of appropriate transfer function estimators, H_1 and H_2.

The Exercises
Exercise 11.1

Objective

To investigate bias errors traceable to insufficient frequency resolution.

Reminder

Analysis resolution: $\Delta f = \dfrac{1}{N\Delta t}$

SDOF system bandwidth: $\mathrm{BW} = 2\varsigma f_0$

with ς the damping ratio, f_0 the undamped natural frequency (Appendix 4.A in Part B).
The normalized random error for $|H|$ is

$$e_{|H|} = \frac{[1 - \gamma_{xy}^2]^{1/2}}{|\gamma_{xy}|\sqrt{2M}}$$

Description

SDOF or 2DOF systems may be chosen. The bandwidth is controlled, via the damping factors. In the analysis, the resolution Δf is controlled via N, the length of time window. Analysis results (blue) are

Discover Signal Processing: An Interactive Guide for Engineers   S. Braun
© 2008 John Wiley & Sons, Ltd

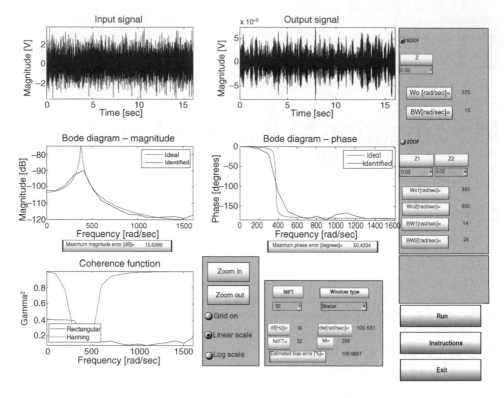

Figure E11.1

shown both for the FRF (magnitude and phase) and the coherence function, theoretical ones are shown in red (Figure E11.1).

A rectangular or Hanning window can be chosen, with results shown in blue and red respectively.

Instructions

- For the SDOF case, check the effect of the parameter NFFT on the transfer function, as a function of the damping ratio z.
- For $z = 0.02$, check the effect of NFFT on the coherence function, using the two types of windows available.
- Check the error magnitude (as printed in the appropriate box) for various values of NFFT, by running the program repeatedly.
- For the 2DOF case, choose $z_1 = 0.02$, $z_2 = 0.05$. Vary NFFT and check the effect on the transfer function accuracy. Repeat for the coherence function, applying both types of windows.

Tasks

First for the SDOF case, it is required to find a rule of thumb for acceptable bias errors, both for the FRF and the coherence function, relating Δf and the bandwidth (both noted on the screen). This is to be done by choosing various combinations of damping and N. The effect of the window on the coherence bias error is also to be investigated.

Next repeat for the 2DOF case. Use the characteristics obtained from the SDOF case to explain the behavior of the coherence function.

Exercise 11.2

Objective

To demonstrate the bias error induced by delay of the output signal.

Reminder

A delay introduces a phase shift proportional to the frequency:

$$\phi = \omega\tau$$

A delay in one of the channels results in a bias error, H_1, being underestimated:

$$H_1 = H_0\left(1 - \frac{\tau}{T}\right)$$

with H_0 the true transfer function, τ the delay and T the signal duration spanning $N\Delta t$.

Description

Basically this corresponds to Exercise 11.1, for an SDOF system with fixed parameters. However, a (controllable) delay can now be introduced between the output and input of the system (Figure E11.2).

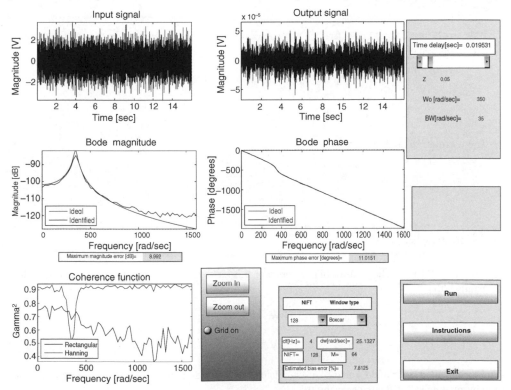

Figure E11.2

Instructions

The correct analysis parameter first needs to be chosen, by determining an appropriate N for the zero delay case. Next investigate the effect of delaying the output.

Tasks

Compute the predicted phase function as a function of the delay. Compare the predicted and computed phase for N large enough to avoid a bias error. Next introduce a bias error when computing with insufficient frequency resolution, and compare with the predicted bias error.

Exercise 11.3

Objectives

To investigate bias errors quantitatively. To see how a large random error can occur for random signals, and be controlled by the choice of the number of segments averaged. To see the effect of additive noise on the random errors. To experiment with a more complex 2DOF case.

Reminder

Bias errors due to insufficient frequency resolution will cause underestimation of peaks. Needed is a resolution of

$$\Delta f \approx \frac{1}{4} \quad \text{to} \quad \frac{1}{3} \quad \text{of the bandwidth BW}$$

$$\text{with} \quad \Delta f = \frac{1}{N \Delta t}$$

The normalized random error is

$$\varepsilon_H = \frac{[1-\gamma^2]^{0.5}}{|\gamma|(2M)^{0.5}}$$

Instructions

First the SDOF case is checked for bias errors. For this case, the sampling interval is 2.5 msec. Both the bandwidth

$$\text{BW} = 2*z*f_n$$

$$\Delta f = \frac{1}{N \Delta t}$$

can be controlled (via z and f_n) as well as calculated, and the bias errors for any case can now be measured by reading selected cursor positions.

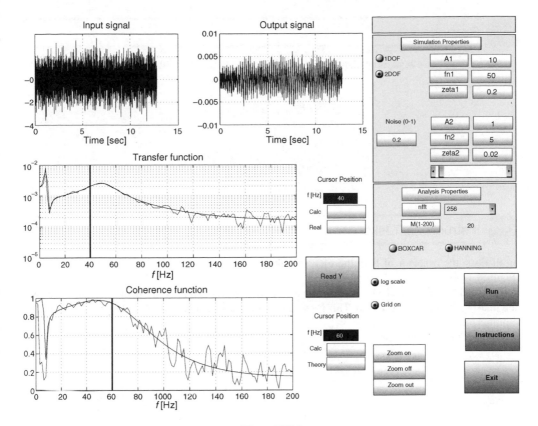

Figure E11.3

Then, with bias errors made negligible by an appropriate choice of NFFT, investigate random errors. Change the number of averages M and the additive noise level. Check the effect of the number of averages and the level of output additive noise. Check the effect of using different windows.

For the 2DOF (Figure E11.3) choose first: $A_1 = 1, A_2 = 1, fn_1 = 22, fn_2 = 30$, zeta$_1$ = 0.002, zeta$_2$ = 0.01, noise = 0, NFFT = 512, M = 200. Check bias errors, random errors and the coherence function for various NFFT and M.

Tasks

Demonstrate the bias and random error for the SDOF case. Summarize the effect of the additive noise, the window type and the frequency resolution on the coherence function.

For the 2DOF case, explain the coherence function for those cases using the parameters as required in the above exercise instructions. Discuss the possibility of different bias errors in different frequency regions.

Exercise 11.4

Objectives

To investigate the performance of two estimators of the FRF, for cases of additive random noises.

Reminders

For additive noise m at input and n at output, estimators H_1 and H_2 will compute:

$$H_1 = H_0 \frac{1}{1 + \dfrac{S_{mm}}{S_{xx}}}$$

$$H_2 = \frac{H_1}{\gamma^2} = H_0 \left(1 + \frac{S_{nn}}{S_{yy}} \right)$$

Instructions/Tasks

In all tests, the behavior of both the FRF magnitude and the coherence function is to be analyzed. (Figure E11.4).

H_1 is used first. For the SDOF case investigate the effect on input noise only, then of output noise only and finally of both noises is tested. This is repeated for H_2, and finally both H_1 and H_2 can be inspected simultaneously.

Repeat for the 2DOF case.

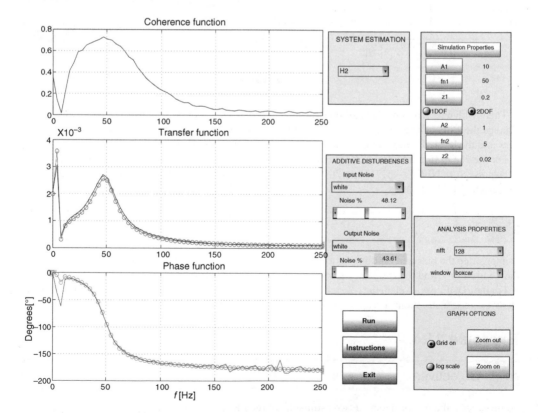

Figure E11.4

Solutions and Summaries

Exercise 11.1

The sampling interval: from the case NFFT = 32, Δf = 16, we get Δt = $1/2^5 2^4$ = 2^{-9} [sec]. The ratio of the bandwidth to the analysis resolution dictates the bias error. This itself underestimates peaks. For the SDOF case, this can be seen for example by choosing z = 0.02 and 0.05 respectively, and changing NFFT in the range 32 ... 512. A ratio of at least 3 seems indicated.

For z = 0.02, the 3 dB bandwidth is $15/(2\pi)$=2.4 (Hz), and for NFFT = 256, Δf = 2, with an unacceptable bias error (Figure E11.5a). Choosing NFFT = 1024 will give Δf = 0.5, with acceptable results (Figure E11.5b).

Running the program repeatedly shows the variability of the estimated transfer function, and hence of the error magnitude. For z = 0.02 and NFFT = 128, this is in the range of 2–4 dB (based on the screen printout). This error increases for higher values of NFFT as M, the number of averages used in computing H, decreases. Hence we note again (as in Chapter 7), the trade-off between bias and random errors for fixed duration data.

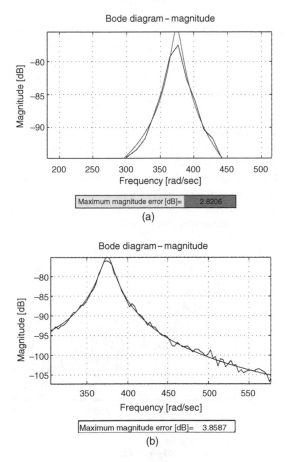

Figure E11.5

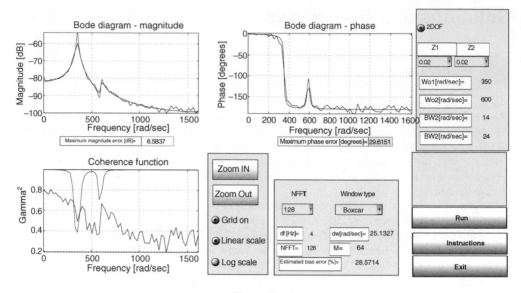

Figure E11.6

For insufficient resolution (too small NFFT), bias errors occur for the estimated coherence function as well. The character of the bias error as a function of frequency depends on the window used: shown is a gradual decrease for a rectangular window, with a sharp minimum around resonance. *It could then be used to identify a resonance!* With a Hanning widow, the bias error function shows only a negative gradient, without the phenomenon of the dip at resonance.

Similar characteristics can be seen for the 2DOF case, however some care is needed for the interpretations. Note that bias error may occur in none, both or only one region (Figure E11.6), depending on the existing bandwidths for each region.

The coherence function is more complex (Figure E11.7), a case with NFFT = 512. A minimum exists at those frequencies where the output is negligible, i.e. at the minima of the FRF (around 580 Hz). A minimum is observed around the first resonant frequency when a Hanning window is applied (see the SDOF case). However, no minimum is observed for the second resonant frequency, where the bias error is less, as the bandwidth is larger. Due to the random error we get variations in the plots for each consecutive run. However, this minimum is always evident.

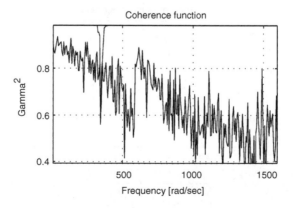

Figure E11.7

Exercise 11.2

For zero delay, bias errors will be due to insufficient Δf. Choosing NFFT = 512 shows a very small bias, and NFFT = 1024 results in a negligible one.

For a delay of $\tau = 0.19922$, at $\omega = 26.2$, the predicted phase shift is

$$\omega\tau = 26.2*0.19922 = 5.2196 \text{ [rad]}$$

This corresponds to the computed result (Figure E11.8), showing (after zooming in) a phase shift of 300 degrees, corresponding to 5.236 [rad].

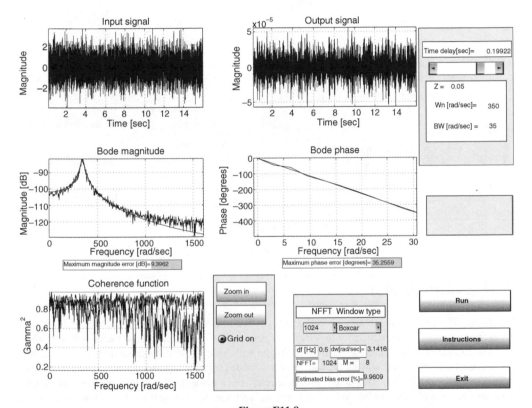

Figure E11.8

Bias error occurs in the computed FRF. For the case shown, choosing NFFT = 256 gives a frequency spacing of $\Delta f = 2$, and a signal duration in the analysis window of $T = 1/\Delta f = 0.5$ [sec]. With $\tau = 0.19922$, the predicted bias error is

$$20*\log(10)\left[\frac{\tau}{T}\right] = -8\,\text{dB}$$

in reasonable agreement with the result shown, after zooming, in Figure E11.9. The delay also results in a bias error in the coherence function, where the choice of window, for the latter, affects the shape (see Exercise 11.1).

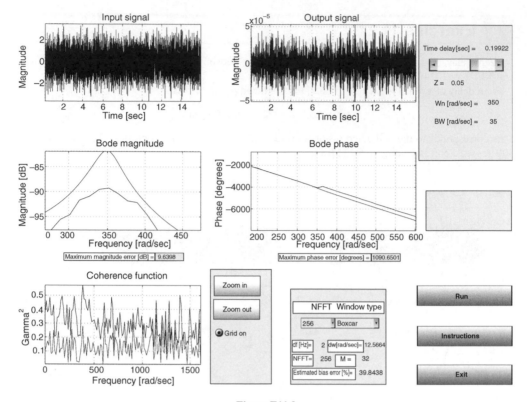

Figure E11.9

Exercise 11.3

For the SDOF case, zero noise, zeta$_1$ = 0.01 and fn_1 = 5, a low NFFT (say 32) will cause a bias error in $|H|$ at the resonance frequency (close to 40%). It is necessary to zoom in and use the cursor readout to read the calculated values, and the result will change with each run due to the random error. The bias in the coherence function is noticeable, with very low values for a rectangular window, and a strong dip at resonance for the Hanning window. This has already been evident in previous exercises. Increasing zeta$_1$ and/or fn_1 will increase the system's bandwidth, and hence decrease these bias errors. Increasing NFFT decreases these bias errors, and with NFFT = 4096, the error in $|H|$ is practically nonexistent, and the coherence function at resonance is above 0.95 with a rectangular window (Figure E11.10). However, the coherence function drops rapidly at other frequencies, where the output power is reduced significantly. With a Hanning window, however, the drop of the coherence function at resonance is obvious, but it is close to 1 in all other regions. Adding output noise (say n = 0.4) has little effect on $|H|$.

　　The coherence function with a rectangular window shows a decreased value, as it is sensitive to output noise, and is obviously close to zero at frequencies distant from the resonance where the coherent output is negligible. With a Hanning window, we note again the sharp minimum at resonance, and the further drop in regions where the coherent output noise is negligible (Figure E11.11). Repeating with lower values of M, the random error becomes noticeable in both $|H|$ and the coherence function.

　　For the 2DOF case, with the required parameter values, we note how the bias error caused by insufficient NFFT depends on the frequency region (Figure E11.12). With two different bandwidths at the

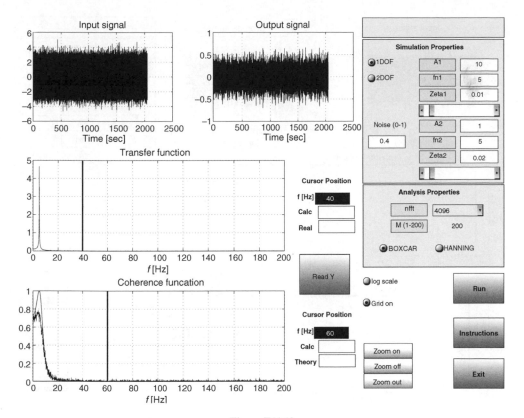

Figure E11.10

two resonances, the bias error is very small in the region with large bandwidth, and large in the low frequency region with the small bandwidth. Thus the bias error may have to be assessed separately for different frequency regions! Also a zero of the system can be noted at 15.2 Hz, and the coherence function, shown above for a rectangular window (with zero coherent output), should always be zero there.

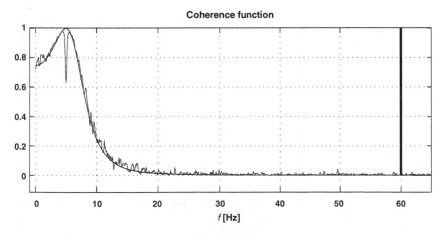

Figure E11.11

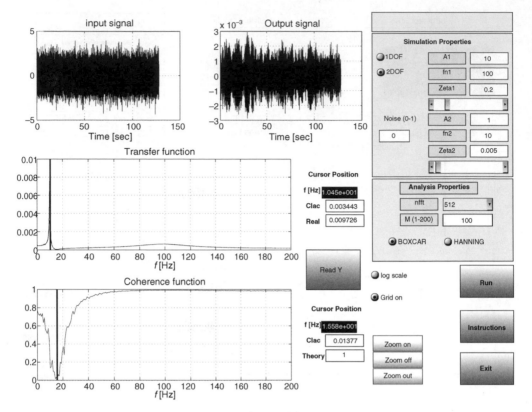

Figure E11.12

With a Hanning window there can be an additional minimum at resonances, depending on the bias error as dictated by NFFT (Figure E11.13).

Adding noise and using smaller values of M will show the effect of the random error (Figure E11.14). Using $M = 1$ results in a computed coherence function equal to 1 at all frequencies, whatever the true (physical) coherence.

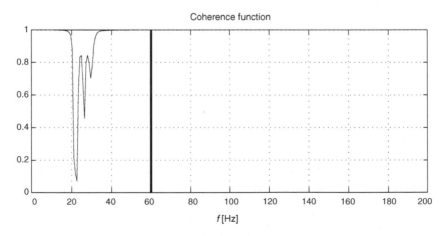

Figure E11.13

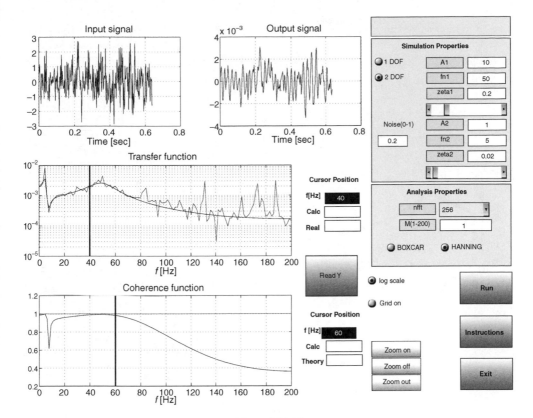

Figure E11.14

Exercise 11.4

For the noise-free SDOF system, we need NFFT = 1024 for an acceptable bias error. The coherence function is close to unity with a rectangular window, and shows the dip at resonance with a Hanning window.

Adding white noise at the output has no effect on $|H_1|$, but overestimates $|H_2|$, adding noise at the input underestimates $|H_1|$ but has no effect on $|H_2|$. The coherence function is reduced in all cases by the additive noises.

The 2DOF case shows similar effects. There is, however, a system zero at around 7 Hz, seen in all the computed functions (Figure E11.15). Using a Hanning window, we now observe two minima, one at the low frequency resonance (5 Hz) and one at the frequency of the system's zero (7 Hz) (see Figure E11.16).

Additive noises at the input and output affect the transfer function estimators and the coherence functions as before.

It is instructive to check the effect of adding low passed noises. The bias error (over-or underestimated, depending on the location of the additive noise and the choice of estimators) as well as the coherence function, are affected only in the low frequency region, whereas the

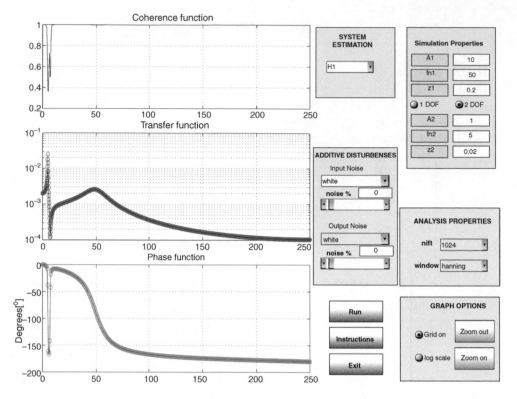

Figure E11.15

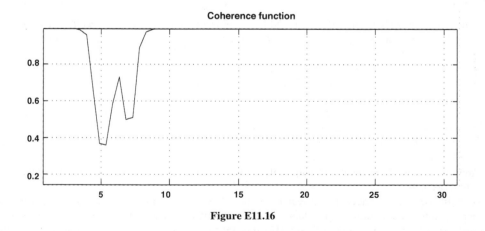

Figure E11.16

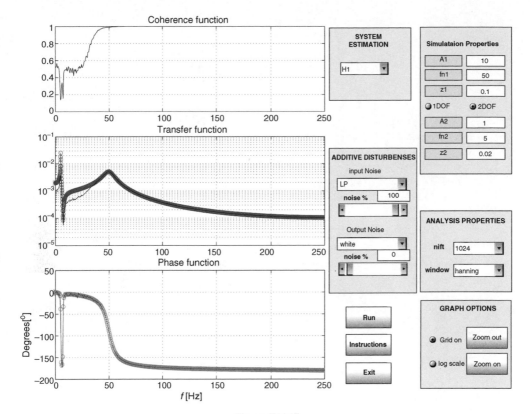

Figure E11.17

higher frequencies are unaffected (Figure E11.17). The coherence function may thus indicate acceptable results in specific frequency regions, and unacceptable ones in others. *Basically we have a powerful computational tool able to accept/reject results differently in various frequency ranges!*

12

Model-based Signal Processing

Overview

These exercises are intended as an introduction to model-based signal processing, a topic usually meriting a dedicated book. Exercise 12.1 deals with spectral analysis, and its comparison to FFT-based estimations. Exercise 12.2 shows one application example, dealing with detecting *changes* in signal. Some insight into the basic problem of model order determination is introduced via Exercise 12.3.

The Exercises
Exercise 12.1

Objective

To compare a parametric and nonparametric approach to spectral analysis.

Reminder

- Model-based PSD result from a parametric analysis.
- The lower the model order, the smoother (smaller random error) the PSD.
- Unless known via external knowledge, the model order must be determined by a criterion based on the data itself.

Description

Spectral analysis is performed via an autoregressive (AR) model, and the analysis properties consist in choosing a model order p. Two signals can be chosen, one consisting of experimentally acquired vibrations of a clamped beam, another being a noisy sine signal. The signal is shown in the upper plot (Figure E12.1). The middle plot shows two superimposed PSDs – the model-based (red) and Fourier-based (black) ones. The lower plot shows the final prediction error (FPE) criterion computed for the complete range of available orders (blue), while a red circle shows the FPE for the chosen model order. Log, grid and zoom options are also available.

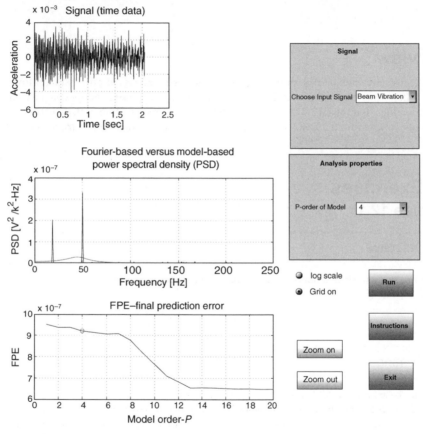

Figure E12.1

Instructions

Choose the beam vibration signal, compute the model-based PSD for varying model orders.
   Repeat for the noisy sine signal.

Tasks

Compare the PSDs as computed via the nonparametric (Fourier) and parametric approach. Explain the effect of choosing various model orders and check the effectiveness of choosing them via the available criterion. (Both the linear and log display options should be used.)

Exercise 12.2

Objective

To demonstrate one application of model-based processing to the diagnostics and monitoring of systems.

Reminder

Having knowledge of a model H_0 from a reference data we may test the residual obtained by using this model and the tested data

- If $H = H_0$, i.e. the residual has white noise properties;
- Else $H = H_1$, i.e. a change occurred.

Sequential parameter identifications can monitor the time evolution of model parameters.

Description

This exercise attempts to introduce a model-based method geared to detecting changes occurring in a signal. A reference signal is used first, and an AR model fitted to it. The model is then used in order to test a second signal, as to whether it belongs to the same class.

Two signals, A and B, are available, shown in the upper plots (Figure E12.2). The reference signal is chosen as one of them, a second signal (the same if so desired) can be chosen for classification.

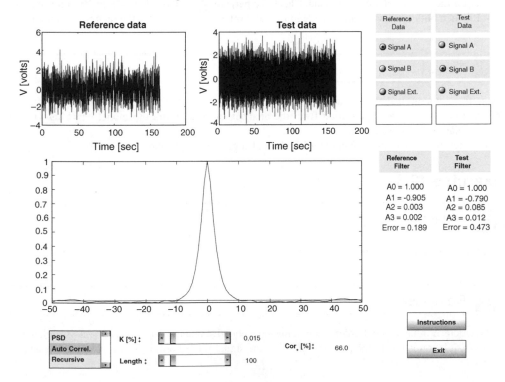

Figure E12.2

The lower left pop-up enables us to choose the analysis method. Choosing the autocorrelation option will generate two AR models for the signals, the numerical parameters of which are shown in the right-hand boxes. The model based classification is then undertaken, the residual of the reference model being inverse filtered by the second model. The autocorrelation of the residual of resulting from this second step is then shown in the middle plot. A horizontal red line shows the upper confidence limit for the autocorrelation of white noise beyond the $\tau = 0$ point (where the power of white noise is concentrated). The confidence level is controlled via the lower slide K, and the percentage of the correlation function below this chosen value is shown in the Cor[%] box. The range for which this autocorrelation can be checked visually can be controlled via the lower Length slider. When changing L, the correlation command must be activated again to obtain the correct horizontal axis. Choosing the PSD option shows the PSD of the two chosen signals. Choosing the recursive action will first append the two chosen signals, generating one longer signal. The AR parameters are then computed recursively, and their evolution depicted in the lower plot. However, this option will only work if the Matlab identification toolbox is available.

Instructions

Apply the PSD command for

- A reference and A test data
- B reference and B test data
- B reference and A test data.

Repeat for the autocorrelation command. Check the percentage of points below $K = 1\%$, for ranges of $L = 10$ when both signals are the same, and $L = 500$ otherwise. Repeat for the recursive command.

Tasks

From the autocorrelation test, decide whether signals A and B differ. Summarize the different information and possible application from the analysis of spectra and the model checking via the autocorrelation of residuals.

Interpret the result of the sequential methods, and suggest an application.

Exercise 12.3

Objective

To discover the effect of overestimating model order.

Reminder

Overdetermining a model order is sometimes beneficial. In addition to generating noise-related poles, it is found that in the presence of noise, the signal-related poles will get closer to the true ones.

Description

The program generates a noisy transient response of an underdamped system. A SDOF or 2DOF system can be chosen, with true orders of 2 and 4 respectively. The system true order can be chosen, as well as the noise ratio. An autoregressive moving average (ARMA) model is fitted via the Prony method, and the poles computed first in the Z plane, and then converted to the S plane via $z = \exp(j\omega dt)$, see also Equation (4.10a).The model's estimated order can be chosen.

   The upper plots in Figure E12.3 shows the generated impulse response. The lower left and right plots show the poles in the S and Z domain respectively. The true poles are shown in red circles, the estimated ones in blue. The 'Stepping animation' option shows a dynamic display, updating the estimated poles as the model order is stepped from 1 to the chosen one. First a fast dynamic display steps through all the orders, then a slower rate enables us to follow each updating.

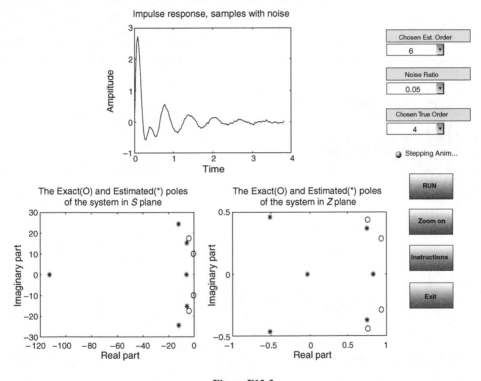

Figure E12.3

Instructions

1. Choose equal true and estimated orders. Run the program for various noise levels.
2. Repeat with increasing values of the estimated model order.

Tasks

- Determine the effect of the noise on the accuracy of estimating natural frequencies.
- Summarize the effect of order overestimation. Check the stability of noise-induced poles.

Solutions and Summaries

Exercise 12.1

For the beam vibration, for $p = 8$, we get results shown in (Figure E12.4), or zooming in and using a linear scale we get results shown in (Figure E12.5). The PSD is very smooth, being based on six parameters only, and only one peak is identified, with no correspondence to the FFT-based PSD.

The FPE criterion indicates an order of $p = 14$. For this order we get results shown in (Figure E12.6). Or, zooming in and using a linear scale, we get results shown in Figure 12.7, and two major peaks corresponding to the FFT-based PSD are identified. A higher model order is indicated if more peaks need to be identified. However, looking at the FFT-based PSD circled area (in the log scale display), it seems impossible to state with any certainty that the peak appearing there is a 'true' peak, or is just due to the random error (see Chapter 7). External knowledge, not just the data, would be needed to make such a statement. Of course, with more data, segment averaging could have been used to decrease the random

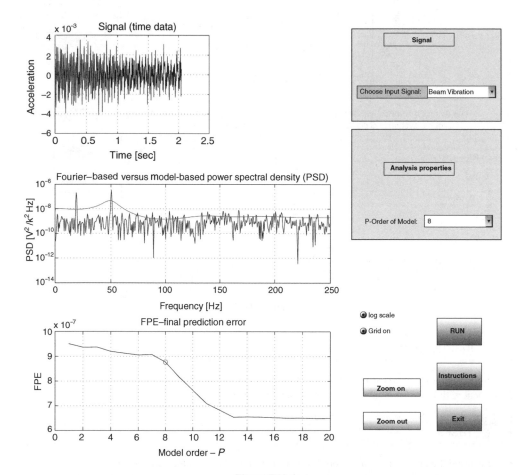

Figure E12.4

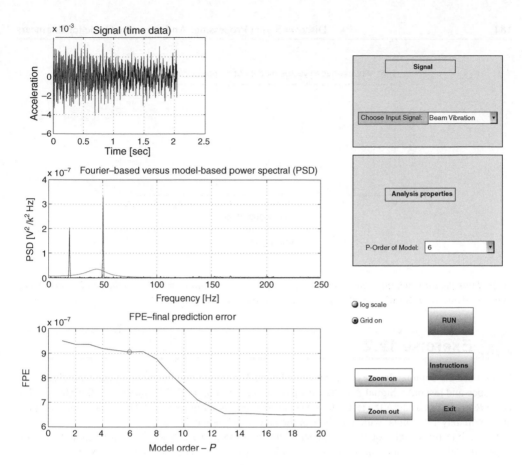

Figure E12.5

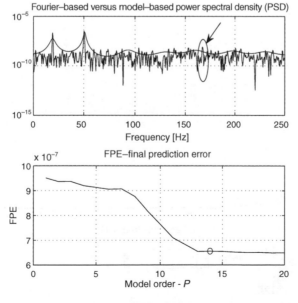

Figure E12.6

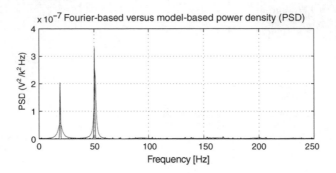

Figure E12.7

error. *Thus deciding whether the peak has a physical interpretation, or is due to the random error, is a subjective decision!* Similar results are obtained for the noisy sine case.

Exercise 12.2

Using the PSD commands, the character of each signal is recognized by choosing the same signal as a reference and test ones. Signal A is thus low pass (as seen in Figure E12.8), and signal B high pass. For B the reference data and A the test data, we get results shown in Figure E12.9. The autocorrelation spans approximately 150 points, with only 20% of the autocorrelation below 1%.

For both tested signals equal, we get results shown in Figure E12.10. The autocorrelation then spans approximately two points. The test certainly indicates the equivalence of both data tested.

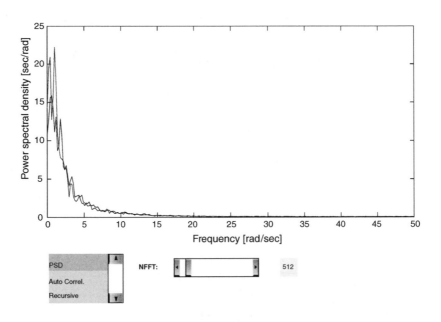

Figure E12.8

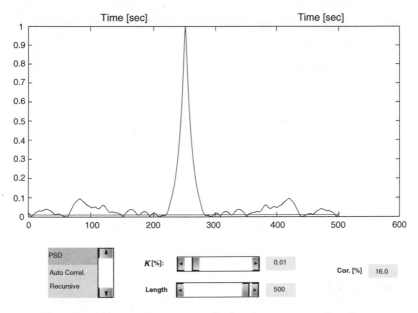

Figure E12.9

Actually we would have 'guessed' that the two signals were different, just from visual inspection of their time history, a subjective evaluation of course. The present test, however, can be considered more objective, enabling us to set a threshold (say the minimum autocorrelation width) for a statistical decision.

The sequential test gives the result shown in Figure E12.11 when the A is appended to B. The abrupt change of all parameters after 4000 data points indicates a change in the signal characteristics. Thus a monitoring scheme, geared to detect changes, is possible.

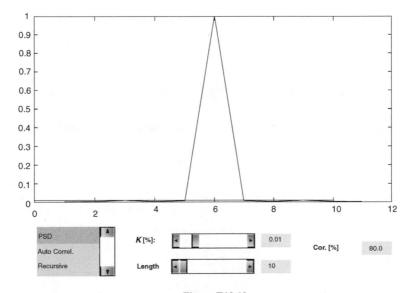

Figure E12.10

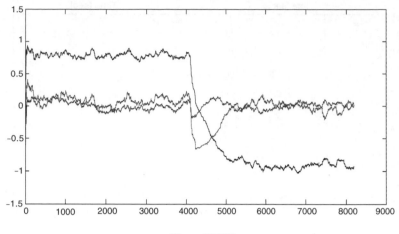

Figure E12.11

Exercise 12.3

For a noise-free system, choosing an estimated order of 4, equal to the true order, results in a perfect identification of poles (Figure E12.12). However, when overestimating the order, to be larger than the

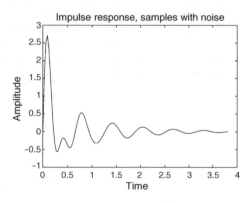

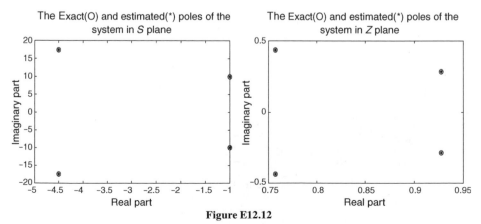

Figure E12.12

true order, or by having a noisy signal, additional erratically located poles occur, and the identified poles are inaccurate. For true order = 4, estimated 6, zero noise, we get results shown in Figure E12.13 And for true order = 4, estimated −4 and noisy signal, we get results shown in Figure E12.14.

Repeating with a significant overestimation, we note the following behavior: while increasing the orders, some estimated poles converge to the true poles, whereas the others vary erratically. The case of $p = 20$ is shown in Figure E12.15. *Thus a possibility of identifying true poles can be based on*

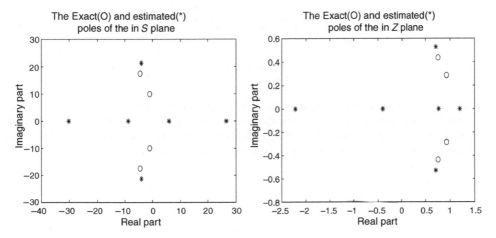

Figure E12.13

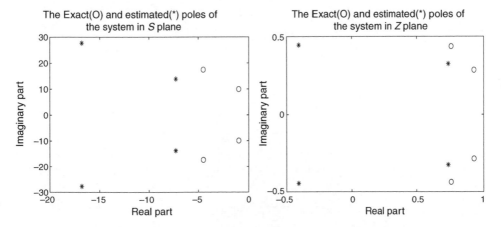

Figure E12.14

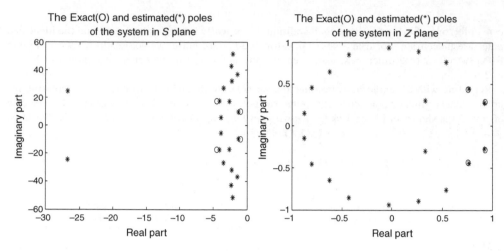

Figure E12.15

sequentially increasing the order (this is sometimes called a stabilization test) and observing poles which converge to a specific value. This is easily seen by choosing a noisy signal with true order 4, an estimated order of 20, and running the stepping animation program. Once a quick movie is finished, a slower display shows how some identified poles converge to the true ones.

13

Diagnostic Applications for Rotating Machines

Overview

Application of the techniques exercised in previous chapters is presented here, addressing diagnostic problems typical to rotating machinery elements. The correct choice of analysis parameters, based on some physical understanding of the systems, is evident in some of these examples. Modulation effects (covered in Chapter 3) are often noticeable. Experimentally acquired data is mostly used, and the difference between ideal theoretical assumptions and real life situations will be easily grasped when attempting to perform the required tasks.

Exercise 13.1 and 13.2 deal with roller bearings diagnostics, the techniques involving filtering, spectral analysis and envelope detection. Exercise 13.3 deals with gear diagnostics, stressing the need to match the spectral analysis parameters to the physical parameters of the system investigated. The existence of modulation phenomena (Chapter 3) and the effect of structural parameters (based on transfer function descriptions as per Chapter 11) are tested in Exercise 13.4. Insights based on time–frequency analysis (Chapter 9) are shown via Exercise 13.5.

The Exercises
Exercise 13.1

Objective

To apply filtering preprocessing to a signal which shows impacts generated by a bearing fault. The full process of bearing fault identification, which could benefit from such a preprocessing, will be explored in the next exercise.

Reminder

A filtering process attempts to improve the signal to noise ratio for the (fault-generated) signal component of interest.

Description

The basic filtering operations available here were already encountered in Exercise 5.2. The signal to be filtered is shown in the upper left plot in Figure E13.1, its spectrum in the upper right one. The filtered signal and its spectrum appear in the lower plots. The filter's design include choice of filter and its order, and the critical frequencies (via cursors). Linear phase filering is available too, but it is based on a method whereby the filter order is twice that seen in the 'Order' box.

The procedure is as follows: choose the signal and then 'Run'. Next choose the filter (class, type, order) and apply 'Design filter', then 'Apply filter'.

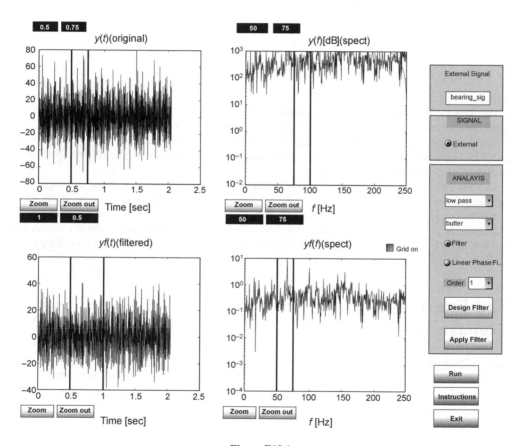

Figure E13.1

Instructions

Choose the external signal named bearing_sig. Apply various filters (low pass, band pass and high pass) of different orders, and visually inspect the time signals. At the end try the following: Butterworth high pass filter, order 6.

Tasks

Show that filtering can improve the recognition of the impulsive shocks generated by a defective bearing with a localized fault. Determine whether the necessary filter parameters can be determined in a simple manner.

Exercise 13.2

Objective

To analyze and interpret roller bearing vibrations. To compare spectral analysis in conjunction with demodulation based on filtering and envelope detection.

Reminder

- Outer race fault frequency components: kf_0, $k = 1, 2, 3...$
- Inner race fault frequency components: $kf_1 \pm qf_r$ $k = 1, 2, 3..., q = 1,2,3...$
- The power of the measured signal $X(f)$ is concentrated around the resonant frequencies of the system. That of the envelope $X_{env}(f)$ is shifted to the low frequency region; however, it retains the pertinent patterns.

Description

The data to be analyze±prises the following cases: a no fault bearing (Nof), a bearing with a localized fault in the inner race (IR) and one with a localized fault in the outer race (OR). The shaft on which the bearing is mounted rotates at 801 rpm, the predicted frequencies generated by the inner and outer race faults are 94.88 and 65.32 Hz respectively. The sampling frequency equals 12 800 Hz.

The data to be analyzed is chosen via the upper left menu, clicking on the box with three dots. This opens a dialog box, showing the data files available. Choosing the desired file and then 'Continue', will input the data to the program, which is started via the 'Run' command. The signal is shown in the upper left plot, its spectrum in the upper right. The spectral analysis uses NFFT, which is chosen (as well as the window) via the analysis parameters (Figure E13.2)

Two cursors in the upper right plot will choose critical frequencies for filtering the data. The critical frequencies chosen and the resulting bandwidth are displayed on the right. Zoom in enables us to look at details in any plot, Zoom off will freeze the zoomed-in plot for any subsequent action, for example moving the cursors. The filter is activated by the 'Double click to filter' button, and an envelope is computed. The lower left plot shows the superimposed filtered signal and its envelope, the lower right plot the PSD of the envelope. Cursors are available to read the location of spectral peaks, as well as the separation df of the frequencies chosen by the cursors.

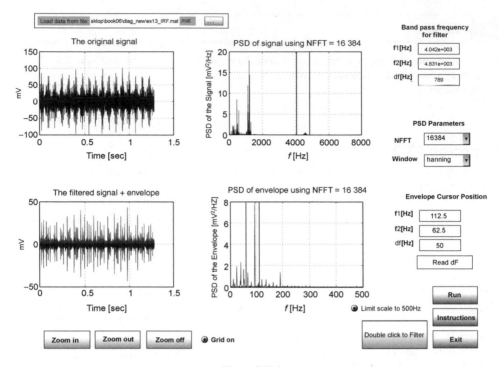

Figure E13.2

Instructions

1. Open the inner race fault signal (IRF), and run the program.
   (a) Filter around 120–850 Hz, and inspect visually the signal (using zoom) and the filtered signal. Identify the peaks in the envelope spectrum.
   (b) Repeat when filtering the 4–5 kHz band.
   (c) Repeat when filtering the 1-2 kHz band. Check also the effect of choosing NFFT.
2. Repeat for the outer race fault (ORF) fault.
3. Repeat for the no fault (NoF) fault.

Tasks

Identify the predicted frequencies for the IR default case. Discuss whether these can be identified in the original and envelope-based spectrum. Discuss the time domain signal shape before and after the filtering operation, and also the results for the different frequency ranges chosen. Compute the minimum necessary NFFT for this task, and explain the results when a smaller value of NFFT is used. Explain the results for the OR and NOF cases.

Exercise 13.3

Objective

To analyze gear vibrations, and compare predicted with measured frequencies.

Reminder

$$f_{\mathrm{m}} = N_1 M_1 = N_2 M_2$$

Meshing frequency harmonics:

$$k f_{\mathrm{m}}, \quad k = 1, 2, 3 \ldots$$

Modulation effects

$$k_1 f_{\mathrm{m}} \pm k_2 N_i$$

$$k_1 = 1, 2, 3 \ldots \quad k_2 = 1, 2, 3 \ldots \quad i = 1 \text{ or } 2$$

Description

Two vibration signals are measured during the progressive evolution of a gear defect. The first one (designated as state 1) is shown in the upper left plot in Figure E13.3, the second, later one (designated

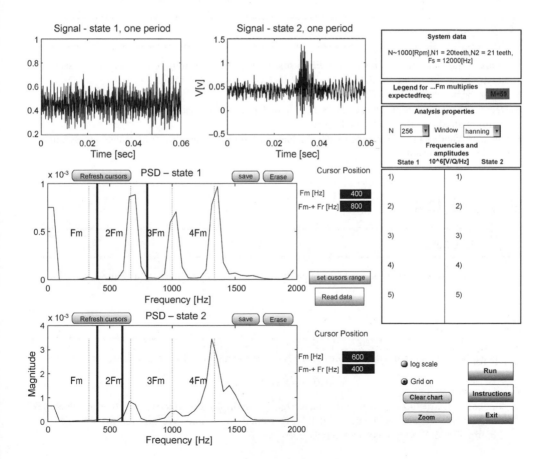

Figure E13.3

as state 2) is shown in the upper right plot. To give additional information, not related to the tasks in this exercise, a tooth breakage occurred later, this being state 2. When the program is called, the first upper plots show the complete available signal. After choosing NFFT they show only the first 60 msec.

The spectral plots are shown in the middle and lower plots, with analysis parameters chosen via the 'Analysis properties' window. Two cursors for each spectral plot can be used to check specific spectral peaks. After zooming (if necessary), clicking on the zoom button again will freeze the figure, enabling us to use the cursors again (as long as they were chosen to be within the zoomed plot). The cursor range can also be controlled via the 'Set cursor range' button, and the sliders which appear when it is activated. Actuating the 'Read data' button transfers the values, as chosen by the cursors, to be transferred to the tables to the right of the plots. Activating then the 'Save' button will transfer the data to the right columns (but a high resolution monitor might be necessary for clear viewing). Up to five readings can be tabulated for each vibration data.

The gear data is given in the upper right box.

Instructions

From the gear data compute the predicted spectral frequencies. From these, estimate the minimum value of N (NFFT) to be used in order to detect these predicted frequencies.

Tasks

- Analyze the data with a value of N (NFFT) smaller than the minimum estimated as necessary (see above instruction), and check that not all frequencies are indeed detected.
- Using the necessary N (NFFT) check whether all predicted frequencies are detected.
- Determine spectral peaks which can be used to differentiate between the gear status state 1 and state 2, and relate this to the time signal's pattern.
- Observe whether the magnitude of the sidebands around the meshing frequencies are symmetric, and discuss the result.

Exercise 13.4

Objective

To understand the spectra of bearing/gear signals as measured on mechanical structures.

Reminder

The response of mechanical structures depends on the excitation and structural frequency response

$$X(f) = S_{ex}(f)H_{struc}(f)$$

Mechanical structures are often characterized by multiple resonances. For rotating machine elements like gears and bearings, default-generated excitations are similar to AM, FM or combined AM–FM signals (see Exercise 3.4).

Description

A harmonic signal of frequency 100 Hz can be modulated both in the AM and the FM mode (simultaneously). The modulating signal's frequency and shape (sine or square) can be chosen, and so can the modulation depth via the modulation index slider. The generated modulated signal passes through a dynamic system which exhibits three resonances. The amplification, frequency and damping ratio relevant to each resonance can be controlled via sliders. White noise can be added to the output.

The upper left plot of Figure E13.4 shows the generated signal, the right one its PSD in the relevant frequency band. The middle right plot shows the frequency response of the system excited by the generated signal, while the left plot shows the additive output noise. The lower left plot shows the system's output. The lower right plot shows the spectrum of the excitation (green), the system's output (blue) and its frequency response function (red).

Three scenarios with preset values for all controllable values are also possible.

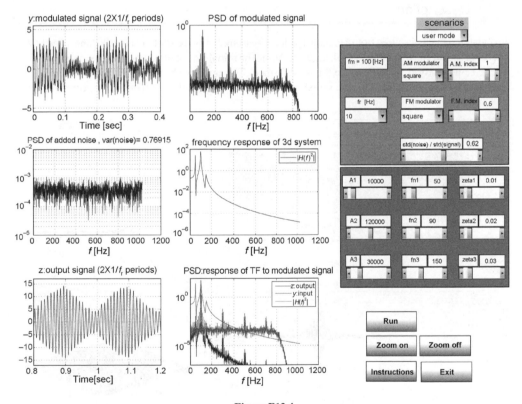

Figure E13.4

Instructions/Tasks

- Run the three scenarios and analyze the results for each case.
- Summarize possible reasons for a possible asymmetry of the sideband amplitudes.

Exercise 13.5

Objective

To suggest a physical explanation for a specific vibration signal acquired from a rotating machine.

Reminder

In real life situations, multiple excitations exist, as any machine will have multiple components, each one generating exciting forces.

Any excitation whose frequency range lies within a resonance region will be highly amplified.

Description

A vibration signal is acquired from a rotating machine, measured from the start-up until reaching a steady state rotational speed. A hypothetical signal, acquired during the initial part of the start-up regime, is analyzed.

The same analysis as in Exercise 9.1 is performed. The upper right plot in Figure E13.5 shows the spectrogram, the lower one the time signal, and the lower right one the 3D spectrogram time frequency

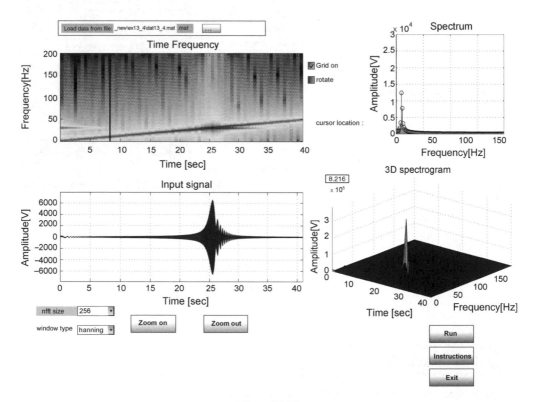

Figure E13.5

plot. This plot can be rotated for any desired viewing plane. A cursor in the upper left plot enables as to choose a reference time, and the spectrum computed for NFFT data points is then shown by the right upper plot. Zooming is available, but zoom off is then necessary if the cursor is to be moved again.

Instructions

Run the program, choosing dat13_5, the vibration signal. Next choose NFFT = 512 and run the program again.

Move the cursor at increments of 5 sec, in the range 0–40 sec Each time note the frequencies as shown by the spectrogram and the upper right spectrum. If necessary, use the zooming option (and grid) to get numerical values for the frequencies shown.

Task

Give a possible explanation for the results obtained.

Solutions and Summaries

Exercise 13.1

The spectrum of the signal (Figure E13.6) shows a complicated structure–the response of a multireso-nance system to an excitation comprising many components. One of those components has the form of periodic impulses, but is mixed with other signal components, whose shape is not known. In this case it is possible to enhance the impulsive component relative to the others by means of filters. The choice of an optimal filter for this purpose would necessitate knowledge concerning the spectrum, the periodic impulses and the other components. As this is not known, an intuitive approach may sometimes be possible. Assuming that short impulsive shocks generate the measured vibrations, these would affect the higher frequency regions. In this case, by concentrating on the higher part of the total spectrum,

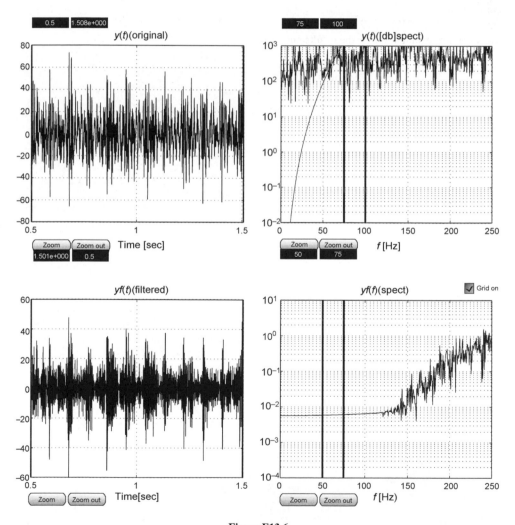

Figure E13.6

the signal of interest (the one showing a repetitive impulsive form) has been enhanced. Note (by visual inspection only) that the impacts are now better defined both in time location and magnitude.

Exercise 13.2

The rotational frequency is $801/60 = 13.35$ Hz. A computational resolution of less than $13.35/3 = 4.45$ Hz will be needed, with NFFT $< F_s/4.45 = 2876$, hence the minimum value is 4096.

For the 120–700 range (Figure E13.7), only the 13.35 Hz component can be identified, reasonably in the envelope-based PSD, barely in the original one. For the frequency range 1–2 KHz (Figure E13.8), both the 13.35, the predicted frequencies $94.88 - 13.35$ and $94.88 + 13.35$, as well as the rotational frequency itself, are clearly seen in the envelope-based PSD. The same sideband pattern exists in the original spectrum, around the filtered frequency band.

Zooming in on the time signals, the narrow band character is noticeable in the plot of the filtered one, an almost constant zero crossing, approximately equal to the reciprocal of the filter's center frequency (Figure E13.9). The (zoomed) upper plot enables a physical interpretation: a sharp transient starts to occur once per rotation, $1/13.35 = 0.074$ sec, and impacts, modulated in amplitude, then occur approximately at intervals of $1/94.5 = 0.0105$ sec. The amplitude modulation is typical for faults on the inner race. Somewhat similar results occur for the range of 4–5 KHz, even if only a minor part of the power is located there (Figure E13.10).

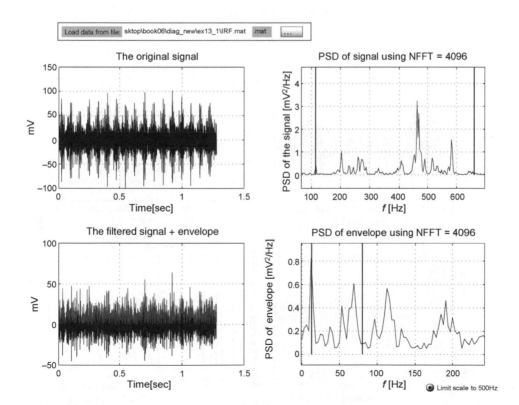

Figure E13.7

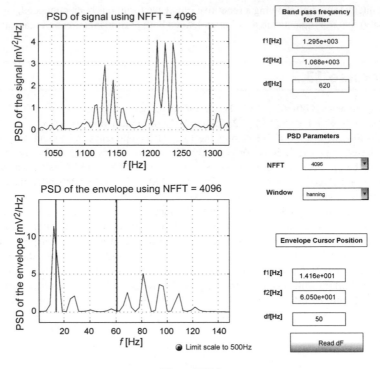

Figure E13.8

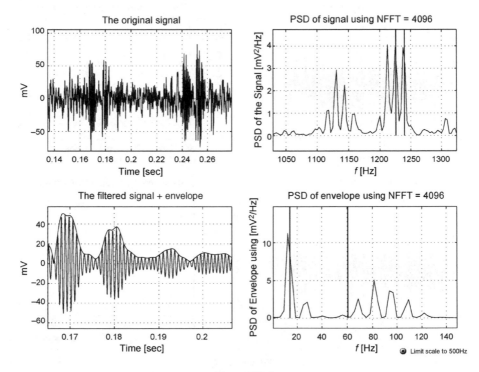

Figure E13.9

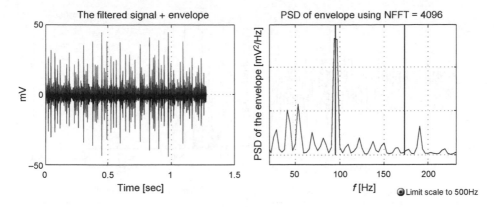

Figure E13.10

The envelope-based detection is still capable of showing the existence of the predicted frequencies. Reducing NFFT to 2048 will show a center frequency but no longer two sidebands; a further reduction will not show any sidebands.

It is now relatively easy to interpret the results for the default in the outer race (Figure E13.11). Depicted is the case for filtering around the 1–2 kHz band. From the (zoomed) time domain plots, it can be noted that the impacts are clearer after filtering, showing that the major power generated by the impacts is in this range. The impacts are separated by 1/65.3 = 0.0153 sec, and there is no amplitude

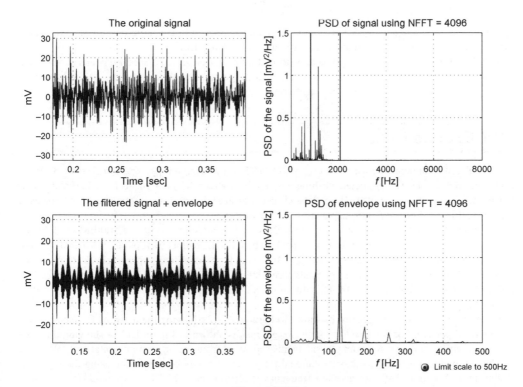

Figure E13.11

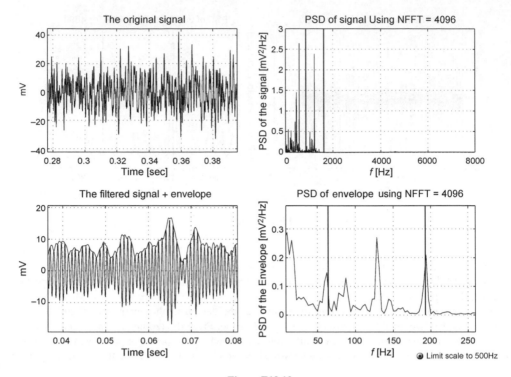

Figure E13.12

modulation. The frequency of 65.3 Hz and its harmonics are seen in envelope-based PSD. Analyzing the no fault case, no predicted frequencies are detected (Figure E13.12).

The filtered signal still shows the typical pattern of narrow band signals. Spectral peaks can be seen, but do not correspond to those predicted by the bearing geometry and kinematics. As the bearing is obviously mounted on a driving system, the frequencies are probably due to other components.

Exercise 13.3

The predicted rotational frequencies are $f_{r1} = 1000/60 = 16.7$ Hz, and $f_{r2} = f_{r1}*20/21 = 15.87$ Hz, a difference of 0.83 Hz. The predicted meshing frequencies are $f_m = N_1*f_{r1} = 20*16.7 = 334$ Hz, and their harmonics $2f_m = 668$, $3f_m = 1002$ Hz. Sidebands occur around the meshing frequencies, at distances of integer multiples of f_{r1} and f_{r2} (16.7, 33.4, 50.1 etc. Hz for f_{r1}).

The highest resolution available occurs for N (NFFT) = 8192, equal to $\Delta f = 12\,000/8192 = 1.46$ Hz. Thus the two corresponding spectral lines, separated by 0.83 Hz, cannot be separated. An analysis considering only one rotational frequency, say $f_r = 16.7$, is thus reasonable in order to check for predicted meshing frequencies. To detect the sidebands, a resolution of $16.7/3 = 5.57$ Hz is indicated, necessitating N (NFFT) = 12 000/5.57 = 2144. The minimum is then N (NFFT) = 2048, with N(NFFT) = 4096 a safer bet.

For N (NFFT) = 1024 we get results shown in Figure E13.13, showing the meshing frequencies, but no sideband pattern. For N (NFFT) = 4096 we get results shown in Figure E13.14 showing the sidebands. Zooming in shows, for example, sidebands separated from the $3f_m$ component at multiples of f_r.

The component related to the basic rotational frequency can only be seen when using the log scale. Zooming in (Figure E13.15), we note low frequency spectral lines at 17, 34 and even 85 Hz.

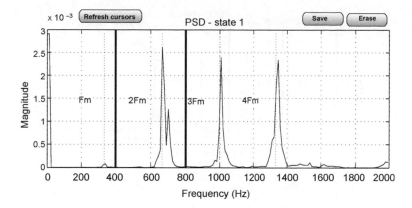

Figure E13.13

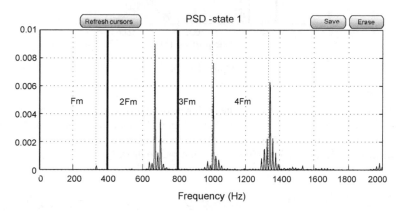

Figure E13.14

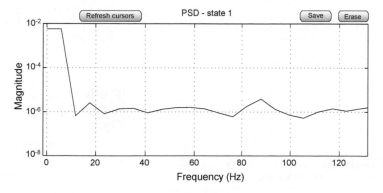

Figure E13.15

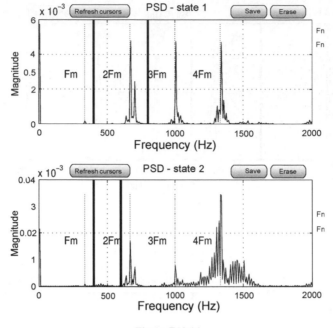

Figure E13.16

Comparing the spectra for state 1 and state 2 shown in Figure E13.16. Zooming in, and reading the meshing frequency and the first sideband level (Figure E13.17), the component at the meshing frequency has grown by approximately a factor of 3 (0.16 as compared to 0.006), the first higher sideband by an approximate factor of 50% (0.0037 as compared to 0.0021).

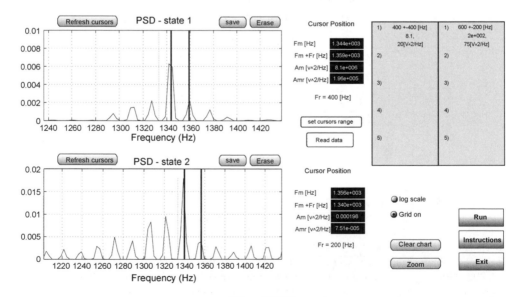

Figure E13.17

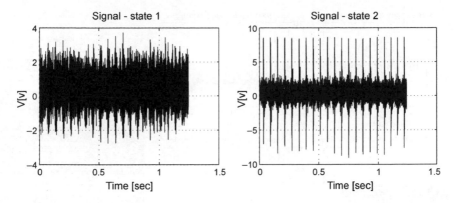

Figure E13.18

It is, however, the increase in the high frequency power which is evident–we note the increased number of significant sidebands around the third meshing frequency (Figure E13.19). This is, of course, traceable to the large transient impacts, generated by the fault, as seen in the raw signal (Figure E13.18). It is easily seen that the sideband amplitudes are not always symmetric.

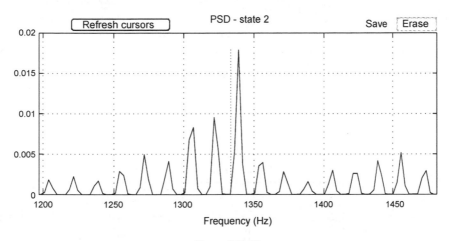

Figure E13.19

Exercise 13.4

Scenario 1 (see Figure E13.20): A square signal is modulated by a square signal, as seen on the upper left plot (Figure E13.20). The spectrum shows multiple sidebands around the harmonic of the 100 Hz carrier. The exciting components in the 100 Hz range are close to one resonance (90 Hz), hence the output signal oscillates around this frequency with some modulation evident. The harmonics of the carrier (100, 200, 300 Hz) are far enough from the resonances, they are strongly attenuated, hence the almost sinusoidal shape of the output signal. Zooming in reveals, however, an asymmetric sideband structure around 90 Hz, due to the variations in the gain for the different sidebands (Figure E13.21).

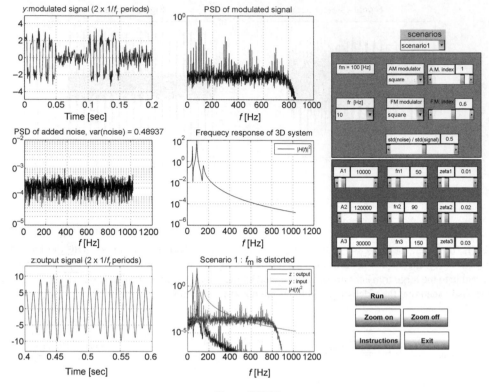

Figure E13.20

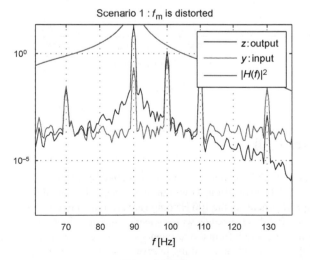

Figure E13.21

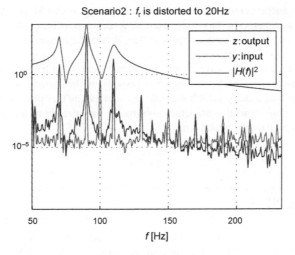

Scenario2 : f_r is distorted to 20Hz

Figure E13.22

Scenario 2 (Figure E13.22): The spectral analysis shows that the carrier of 100 Hz has actually been suppressed, as it almost coincides with a zero of the system. The first set of sidebands (90 and 110 Hz) coincide with two resonances, however the gain is markedly different at these two frequencies. The result is an asymmetry in the sideband amplitudes.

Scenario 3 (Figure E13.23): The third harmonic of the carrier (300 Hz) and the first pair of sidebands (290 and 310 Hz) coincide with resonances. The amplification of the sidebands is large, however. For an amplitude-modulated sinusoidal carrier, modulated at 100% with a harmonic signal, the sidebands would have 25% amplitude of the carrier. Thus the present case would correspond to overmodulation, with a doubling of the envelope frequency (see left plot)–the period of which is now 50 msec–as compared to 100 msec for the modulating period.

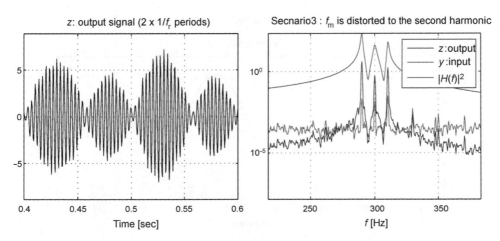

z: output signal (2 x $1/f_r$ periods)   Secnario3 : f_m is distorted to the second harmonic

Figure E13.23

In Exercise 3.4 we encountered a case where the existence of both FM and AM modulation could destroy the sideband symmetry. In the current exercise a second possible cause is recognized: changes of the amplifications encountered by each spectral line, due to the frequency response function of the system which responds to this excitation.

Both conditions can exist for the case of gear vibrations. Both AM (due to eccentric gears) and FM (due to speed fluctuations) may exist. The frequency response of the structure through which the vibration signal propagates will usually exhibit multiple resonances, another possible cause for asymmetry of the sidebands. We now recall that an actual asymmetry was already encountered in Exercise 13.3 which dealt with an actual experimental case.

Exercise 13.5

The time plots show a strongly varying amplitude, starting from zero, growing at an increasing rate to a maximum around 26 sec, and then decaying with an oscillating amplitude to what could be a steady state value (as described for the exercise, a rotating machine is being started up and accelerates to a steady state rotational speed). Zooming in for various times shows an almost harmonic signal, amplitude of 180 and a period of 860 msec (frequency 16 Hz) after 15 sec (Figure E13.24).

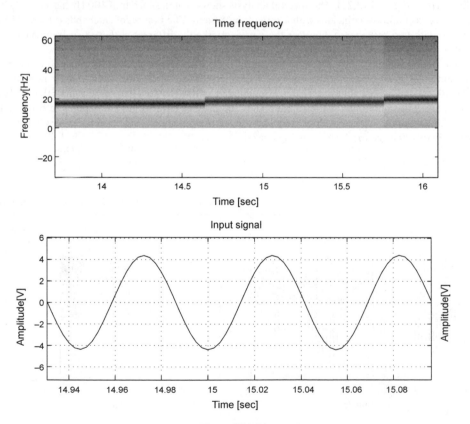

Figure E13.24

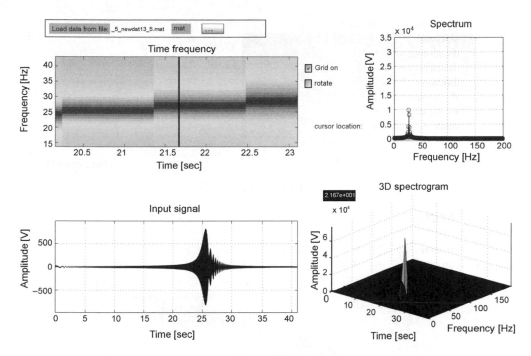

Figure E13.25

Repeating around 22.5 sec (Figure E13.25), we note the significantly increased amplitude, and the period of 30 msec (frequency of 33Hz). Thus the component having a positive slope (increasing frequency) in the spectrogram seems to indicate a vibration directly related to, and tracking, the varying rotational speed, possibly the reaction to a mass imbalance (Figure E13.26). The spectrogram shows an additional component, seemingly of constant frequency, around the start of the signal. Zooming in on the spectrum, this shows a frequency of approximately 30 Hz (see Figure E13.27).

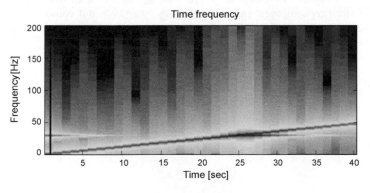

Figure E13.26

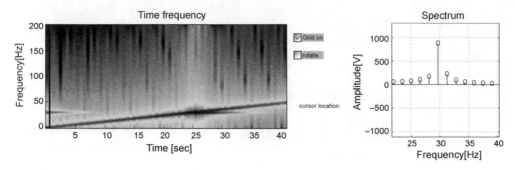

Figure E13.27

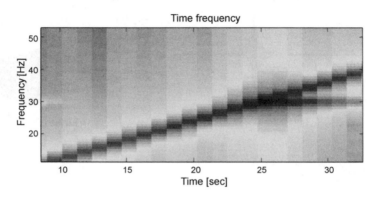

Figure E13.28

Now Analyzing the region where the amplitude has grown sharply, around 26 sec (see Figure E13.28), it can be noted that the component with an increasing frequency crosses the region of 30 Hz. When, during a frequency sweep, a sharp increase in magnitude is encountered for a small frequency region, we often suspect that an exciting frequency coincides with a resonance frequency. Such a component, with this same constant frequency, seems to exist continuously (Figure E13.28). This could also indicate a resonance, excited constantly by additional forces existing in any operating system.

To conclude, a possible reasonable hypothesis would be the following: the system (the monitored rotating machine) exhibits a resonance at 30 Hz. During start-up, until reaching steady state, the frequency of a vibration component tracks the rotating speed. When it coincides with the resonant frequency, a sharp increase in the vibration response occurs, decreasing back to steady state after passing the resonance region.

14

Systems with Delays

Overview

This is the second chapter concerning application of techniques described in earlier chapters. It addresses the detection of propagation delay. Exercise 14.1 utilizes actual acoustical propagating signals measured in a closed pipe, involving analysis of transfer functions and impulse responses. The fact that delayed impulses are recognizable via random excitations and responses, demonstrates how frequency and time domain analysis complement each other. Correlation methods are used in Exercise 14.2, and Cepstrum-based techniques in Exercise 14.3.

The Exercises
Exercise 14.1

Objective

To detect and quantify reflections in the time and frequency domain. To analyze reflections using an impulse response function, obtained via continuous random transmitted/received signals.

Reminder

- Impulse responses can be computed via transfer functions by an inverse FT:

$$h(t) = F^{-1}[H(f)]$$

- Combining a signal and a reflection delayed by τ, results in interferences in the PSD.
- The frequency spacing of the interference is $\Delta f = 1/\tau$.

Description

A 50 cm long pipe filled with air, has acoustically reflecting surfaces at both ends. A piezoelectric transducer is used as a transmitter, applying an acoustic excitation to one end. Acoustic waves then bounce back and forth in this pipe, propagating at the speed of sound in air. Another piezoelectric transducer, acting as a receiver, is located at the other end.

Two types of excitation can be chosen: one consists of a sharp transient impulsive pressure signal. The second consists of a random continuous broadband excitation. The analysis performed depends on the excitation applied. For the impulsive one we have the results shown by Figure E14.1, where the upper plots show the excitation (left) and response (right), the lower left plots show a superimposition of both (left) and a zoomed view (right).

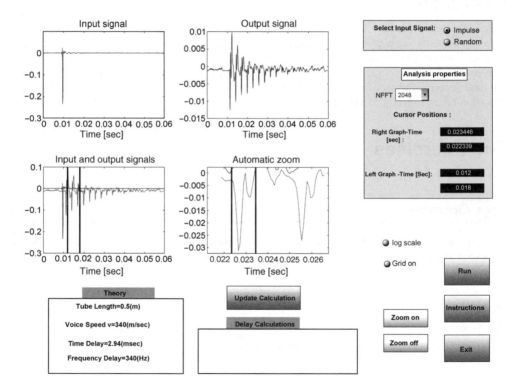

Figure E14.1

For the random excitation (Figure E14.2), we have upper plots with excitation (left) and response (right). The lower left plot shows the transfer function H (left) as computed in the frequency domain (Equation 11.5), and the impulse response computed as the inverse Fourier transform of the transfer function. Cursors are available, enabling us to check the time locations of the lower plots. The locations chosen are printed in the right box, and applying 'Update calculation' shows the time differences (as chosen by the cursors) in the lower middle frame. The theoretical parameters and predictions (based on the speed of propagation), are summarized in the lower left frame.

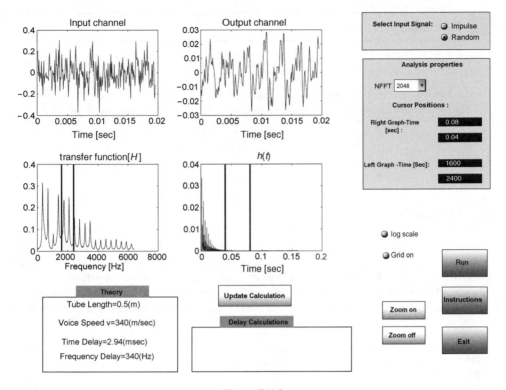

Figure E14.2

Instructions

Choose the impulsive excitation. Using the zoom option, measure the temporal distances between the response impulses.

Choose the random excitation. Measure the temporal distances between the impulse response peaks. Measure the frequency separation between the frequency response peaks.

Tasks

Explain the predicted time delays. Show that these can be checked via the measured signals for both types of excitations. Discuss whether this could have been checked via inspection of the raw time signals only.

For the case of random excitation, discuss the possibility of investigating energy dissipations at the reflection boundaries.

Exercise 14.2

Objective

To see the limitation of the cross-correlation method for the identifying of propagation path delays.

Reminder

$$R_{xy}(\tau) = \sum_q A_q R_{xx}(\tau - \tau_q)$$

and the cross-correlation will peak around τ_q for wideband signals.
The uncertainty theorem relates possible temporal and frequency resolutions, limited by

$$|\Delta\tau| \le \frac{1}{\text{BW}}$$

Description

An acoustic signal is generated and transmitted. The receiver measures the sum of three reflected components, each one reflected differently with a different propagation delay. The cross-correlation between the source and the received signals is used to detect the path propagation delays.

Shown in Figure E14.3 is the computed cross-correlation function. Dashed red vertical lines indicate the delays used in the simulation. These can be chosen via the three sliders on the right. The transmitted

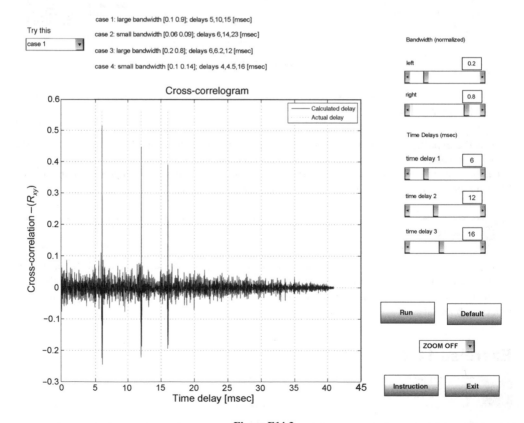

Figure E14.3

signal's bandwidth can be controlled via the two upper sliders. Setting the upper slider to zero generates a low pass signal, otherwise a band pass one is generated.

Four suggested scenarios (bandwidth and delays) can be chosen by the pop-up in the upper left corner. A default case can be chosen by the lower right button.

Instructions/Tasks

Run the four scenarios suggested by the upper left pop-up. Determine the cases for which the delays are clearly identified, and check if this agrees with predictions.

Exercise 14.3

Objective

To analyze and remove echoes by Cepstrum analysis.

Reminder

For a signal and echo, the power Cepstrum

$$\ln|X(\omega)|^2 = \ln|S(\omega)|^2 + \ln[1 + a_0^2 + 2a_0 \cos(\omega\tau)]$$

will show impulses at $\delta(k\tau - t)$ and $\delta(k\tau + t)$, $k = 0, 1, 2$.

For the complex Cepstrum, the complex ln (log) operation permits an inverse operation, and hence to recuperate the time signal. Liftering (filtering) before the inverse operation can remove components.

Description

The transient signal used for this exercise has the form of the impulse response of a SDOF second order system, a decaying oscillating waveform. The damping, natural frequency and signal duration can be controlled. The sampling interval is dictated by the number of points N, which is also controllable. An echo with a controllable delay and amplitude can be added to the signal. Analysis options are power and complex Cepstrum, shown in the middle plot in figure E14.4. For the complex Cepstrum, liftering (filtering) can be applied, with the cursor setting the filter position, and the result shown in the lower plot. The chosen cursor position is printed on screen. To ease the positioning of the cursor, zoom on and off options are available. Noise can be added to the signal via a button.

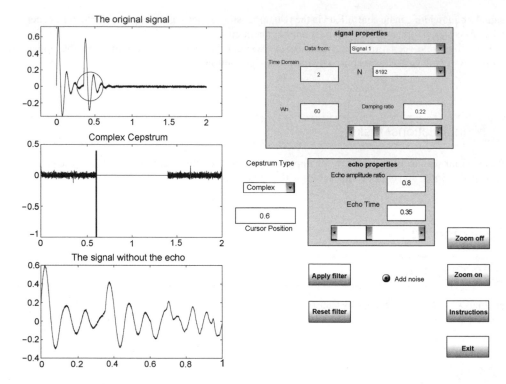

Figure E14.4

Instructions

Use the default parameters which appear when the program is first run. Set the delay to 0.2 sec, and the analysis to power Cepstrum. Determine the location of the spikes in the Cepstrum by using the cursor. Repeat after activating the noise button, and check for the former spikes using the zoom options.

Repeat, while using the complex Cepstrum for the analysis. Set the lifter cursor to 0.2 and apply the filter. Analyze first with, then without, noise.

Repeat for a delay of 0.15, and filter with cursor setting of 0.125.

Tasks

Summarize the information available via the power Cepstrum. Repeat for the complex Cepstrum. Discuss the effect of noise, evaluating its effect on the signal as well the analysis.

Solutions and Summaries

Exercise 14.1

The case of the impulsive excitation is shown by Figure E14.5. The right upper plot shows periodic impulses, of decaying amplitude. Zooming on the lower left plot, the separation between the response impulses is approximately 2.9 msec (Figure E14.6). This is the time delay for a wave to propagate through 1 m. The impulses at the receiving end are thus consecutive arrivals of front to back waves. Checking the separation of the first received impulse from that of the transmitting impulse, the delay is approximately 1.5 msec, the propagation time from the transmitting to the receiving transducer (Figure E14.7). For the random excitation we get results shown in Figure E14.8.

No reflections can be seen by visual inspection of the time signals. However, the indirect method of computing transfer functions and, from it, the impulse response, shows these very clearly. The impulses are separated by 2.9 msec, the same result as obtained from the impulse excitation. Multiple peaks exist

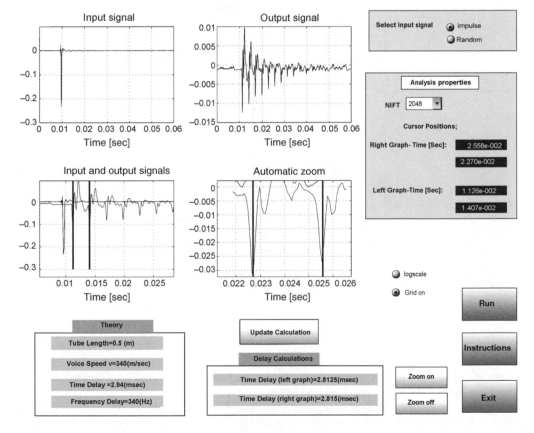

Figure E14.5

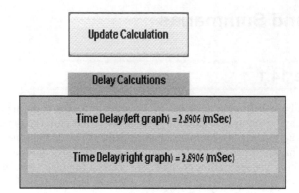

Figure E14.6

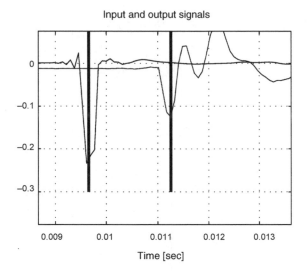

Figure E14.7

in the frequency domain, as predicted by the theory (see Reminder). Spectral peak separation is 340 Hz, the reciprocal of the time delay (Figure E14.9).

The impulse response (obtained by any of the two excitations) shows that the signals are attenuated after each reflection. This certainly indicates an energy loss, and this loss could in principle be quantified and even related to damping/loss models.

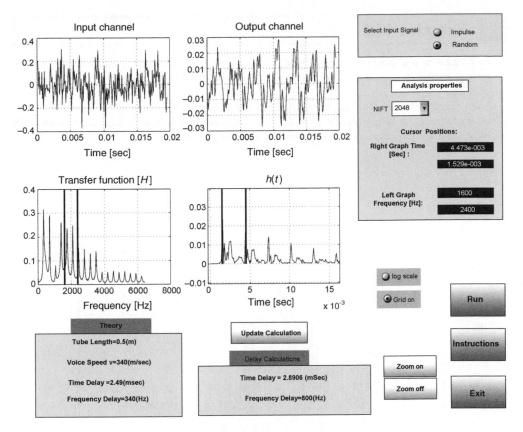

Figure E14.8

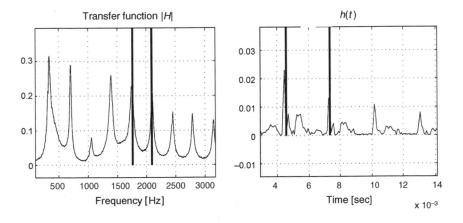

Figure E14.9

Exercise 14.2

Cases 2 and 4 do not show the expected peaks. While case 2 (Figure E14.10) shows some envelope maxima aligned at the expected delays, there would be no way to differentiate those from other spurious maxima.

Case 1 (Figure E14.11a) shows the delays clearly. Case 3 shows these as well, but zooming in may be necessary for an easier visual inspection (Figure E14.11b). The identifications of the delays failed when the bandwidth was too small, as predicted by the uncertainty theorem.

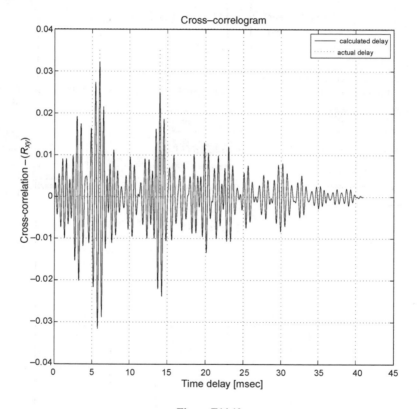

Figure E14.10

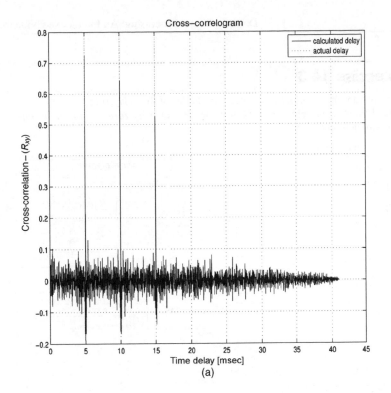

(a)

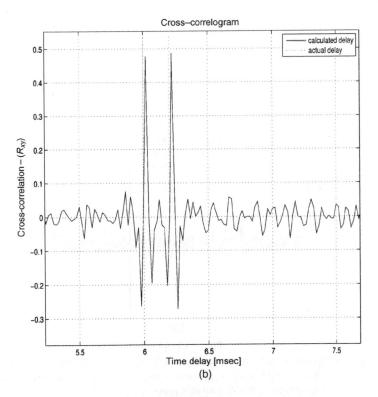

(b)

Figure E14.11

Exercise 14.3

The power Cepstrum shows impulses at integer multiples of the delay, 0.2, 0.4 … msec (Figure E14.12). Adding noise is very detrimental to the Cepstrum. However by zooming in, the impulse can still be identified (Figure E14.13).

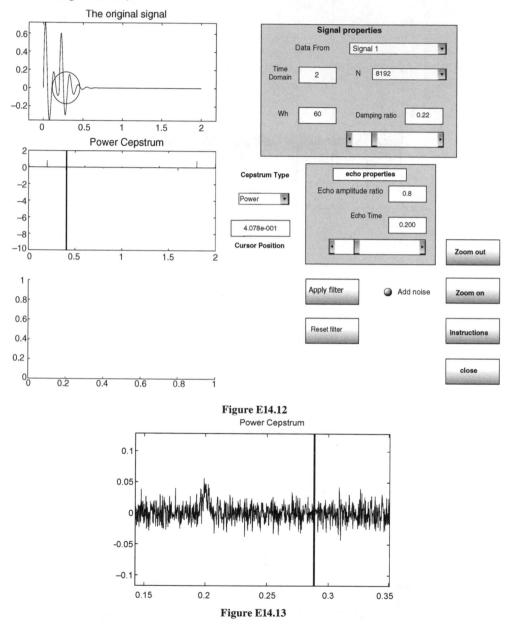

Figure E14.12

Figure E14.13

Using the complex Cepstrum, with a delay of 0.2 and liftering from 0.15, the echo is indeed removed (Figure E14.14). For a delay of 0.15 msec, it becomes difficult to identify the existence of an echo. However, the complex Cepstrum is still effective in removing it (Figure E14.15).

Cepstrum analysis is, however, prone to difficulties in noisy situations. The impulse in the Cepstrum is almost masked by the noise, even if this level of noise barely affects the basic time signal (Figure E14.16). Depending on the noise level, the echo may still be removable (Figure E14.17).

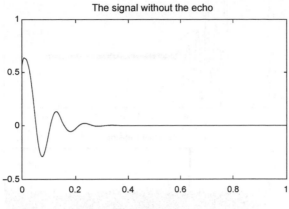

The signal without the echo

Figure E14.14

The original signal

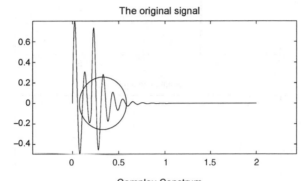

Complex Cepstrum

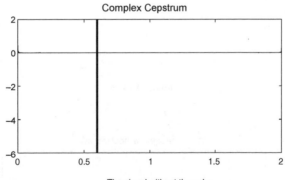

The signal without the echo

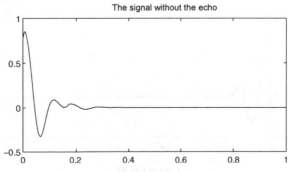

Figure E14.15

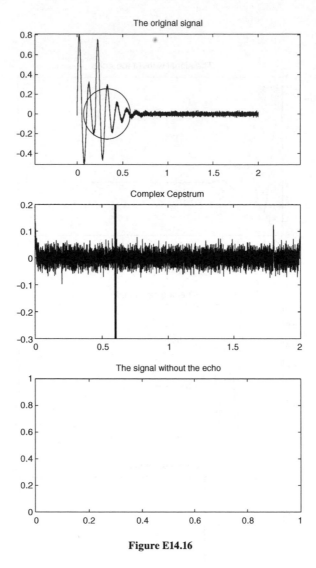

Figure E14.16

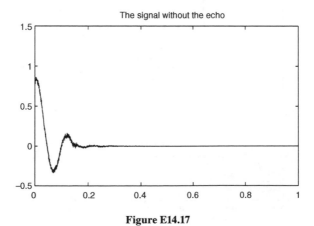

Figure E14.17

Part B

Part B

1

Introduction

1.1 General Objectives

Signal processing is often undertaken for tasks which can be broadly described by the model in Figure T1.1. The available information could consist of the measured output signals only. The objective could be to extract describing functions or parameters, enabling us to monitor, classify or diagnose the operating system.

Extracted parameters can range from the most basic ones like RMS (root mean square), to those sensitive to peaks. Describing functions could be correlation functions, spectral densities, etc., and also parameters directly based on such functions. In other situations, both the excitations and responses are measured. The objective would be identify the system, or sometimes some model parameters only. The excitation could be controlled, as in structural modal testing, or consist of *in situ* exciting ones, for example wind forces acting on existing structures. The identification result could be in the form of a describing function (frequency response function), impulse response or even a model structure.

Many signal processing techniques are geared toward specific situations. However, some of the more basic ones, almost universally applied, would include the following.

1.2 Basic Processing

1.2.1 Filtering

The measured signal(t) is decomposed into components

$$s(t) = \sum_i s_i(t) \tag{1.1}$$

where the various components $s_i(t)$ usually differ in their dynamic character. For example, we could characterize signal components as being 'slow' or 'fast'.

The filtering operation would consist of separating these components, or possibly attenuating (or blocking completely) some of them.

Discover Signal Processing: An Interactive Guide for Engineers   S. Braun
© 2008 John Wiley & Sons, Ltd

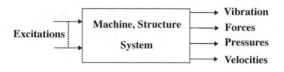

Figure T1.1

1.2.2 Frequency Domain Description

The signal is decomposed into a sum of harmonic components. For periodic functions this is a decomposition into a Fourier series, whereby

$$x(t) = \sum_{k=0}^{\infty} X_k \cos(2\pi k f_0 t + \theta_k) \tag{1.2a}$$

where f_0, the fundamental frequency, is the reciprocal of the signal's period. The function $|X_k(kf_0)|$ is sometimes called a spectrum. In this case of a periodic signal, the spectrum is discrete, composed only of the discrete frequencies kf_0, integer multiples of f_0.

For transient (aperiodic) signals the decomposition becomes continuous in the frequency variable

$$x(t) = \int_{-\infty}^{\infty} X(f) \exp(j2\pi f) df \tag{1.2b}$$

and the spectrum, being continuous, consists of all frequencies in a possibly constricted frequency range. $X(f)$ in Equation (1.2b) is the Fourier transform of $x(t)$, the basic tool for describing signals in the frequency domain.

Many theorems regarding the Fourier transform are extremely helpful. Related to some aspects of this chapter's exercise is the scaling theorem. This states that compressing a signal in the time domain will cause an expansion in the frequency domain (the exact formulation will be presented in Chapter 3). Thus the shorter a transient, the wider the frequency range spanned by its spectrum.

1.3 Why the Frequency Domain?

The use of frequency domain analysis, in many areas of applications, is so predominant that the rationale of using it is often ignored. In what follows is a list summarizing the main possible reasons:

(a) Physical insight is often easier to obtain in the frequency domain, as opposed to the original time domain description of signals and systems. The existence of periodic vibrations occurring with rotating machines is a classical example. The recognition of a constant resonance frequency, excited by a signal, is another one.
(b) The orthogonal properties of Fourier decompositions implies that cross-products of signal components of unequal frequencies are zero. No power is contributed by such cross-products. It is thus possible to investigate independently the contribution of certain frequency regions to the total power or energy. For example, we may try to attenuate acoustic noise in a certain frequency band independently of other frequency regions, and different technological approaches could be tried for different frequency ranges.
(c) Signal patterns of diagnostic significance are often more easily recognized. Small changes, barely affecting the time signature, are often easily detected in the frequency domain representation.

(d) Systems are often modeled as lumped linear systems, and hence described by linear differential equations. Applying a Fourier transform results in sets of algebraic equations. Closed form solutions are readily obtained, and frequency domain descriptions of signals and systems often prevail in introductory textbooks (and this advantage is sometimes cited as the rationale for the Fourier method). While this may be of less interest to real life situations, the closed form solutions are of practical interest, as they are often compared with experimental results. Thus predicted signals (or system) properties are compared with those obtained experimentally, and again the patterns are often easier to interpret in the frequency domain.

(e) The availability of the fast Fourier transform, the now quintessential algorithm for signal processing.

1.4 An Introductory Example

This consists of a base excited mass–dashpot–spring system (Figure T1.2). This is a single degree of freedom system, of second order. Summing the forces acting on the mass results in

$$m\frac{d^2x}{dt^2} + c\left(\frac{dx}{dt} - \frac{dy}{dt}\right) + k(x - y) = 0$$

(1.3)

$$m\frac{d^2x}{dt^2} + c\frac{dx}{dt} + kx = c\frac{dy}{dt} + ky$$

In the frequency domain, this is transformed into an algebraic equation:

$$(-\omega^2 m + j\omega c + k) X (\omega) = (j\omega c + k) Y(\omega)$$

where $X(\omega)$ and $Y(\omega)$ are the Fourier transforms of $x(t)$ and $y(t)$ respectively, and $j = \sqrt{-1}$. This results in

$$Y(\omega) = \frac{j\omega c + k}{-\omega^2 m + j\omega c + k} X(\omega) = H(\omega)X(\omega)$$

(1.4a)

where $H(\omega)$ is the FRF (frequency response function) of the system. $H(\omega)$ can be considered as a frequency-dependent gain factor, applied to each component of $X(\omega)$.

Using generalized parameters, the FRF is

$$H(\Omega) = \frac{1 + 2j\varsigma\Omega}{1 - \Omega^2 + 2j\varsigma\Omega}$$

(1.4b)

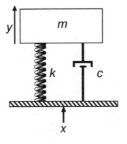

Figure T1.2

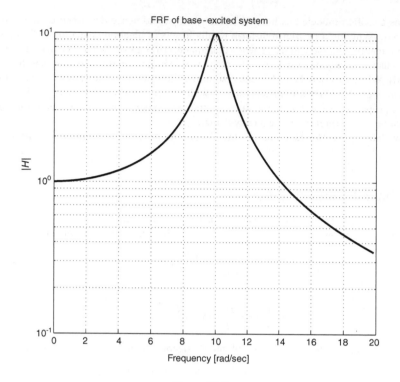

Figure T1.3

with $\Omega = \omega/\omega_n$, $\omega_n = \sqrt{k/m}$, the natural undamped frequency in rad/sec, and $\varsigma = c/2\sqrt{km}$ the damping ratio. For small damping ratios, $|H(\omega)|$ has the form shown in Figure T1.3.

The frequency for which $|H(\omega)|$ is maximum is the resonant frequency $\omega_r = \omega\sqrt{1-2\varsigma^2}$ which for small ς, will be close to ω_n. This model will be used in Exercises 1.1 and 1.2.

2

Introduction to Signals

2.1 Signal Classification

Signal processing approaches will often depend on the signal's properties. Both the characterization as well as analysis methods may depend on the signal structure. The following are some classification possibilities:

- Deterministic vs random
- Transient vs continuous
- Stationary vs nonstationary.

In practice we often encounter combinations of signal types. An example would be a harmonic signal contaminated by random noise.

The exact definition of deterministic versus random may be problematic in theory, but often poses no problem in practice. We may classify a deterministic signal as one whose exact function can be obtained if enough knowledge concerning its generation is available. A random signal, however, can only be described by statistical terms. Impact-generated vibrations would be deterministic, those induced by friction would be random.

Deterministic signals could be periodic or aperiodic (transients). Random signals could be stationary, where their statistical properties are invariant with time, or nonstationary.

Unless we want to use the raw signal, some data reduction is used for its characterization. These will be described according to the signal class. These will be described according to the signal class (Figure T2.1).

2.2 Signal Descriptions

2.2.1 Transient Signals – Energy (Lathi, 1958)

If only one transient is available for analysis, it is usually considered as deterministic. A single description can be based on the energy of the signal

$$E = \int_0^{T_t} x^2(t)\mathrm{d}t \tag{2.1}$$

with T_t the duration. The units of E are not energy per se. However, the energy of any true physical signal (force, velocity, displacement, etc.) will be proportional to E. The type of signal where E is

---

Discover Signal Processing: An Interactive Guide for Engineers   S. Braun
© 2008 John Wiley & Sons, Ltd

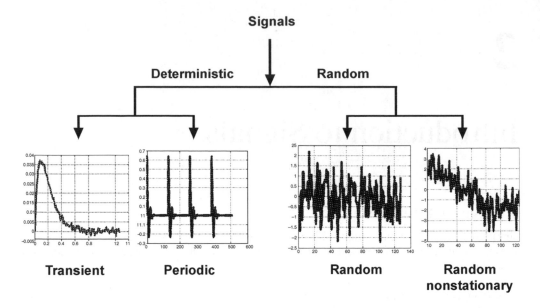

Figure T2.1

finite is called an 'energy signal' and E its energy, with units of V^2-sec, N^2-sec, etc. for voltage, force and other types of signals. The units of voltage are often assumed for $x(t)$, as most measurements are undertaken by electronic instrumentation.

2.2.2 *Continuous Periodic Signals – Power*

A periodic function such that

$$x(t) = x(t + T_p)$$

with T_p the period, is certainly deterministic. The period (or its reciprocal – the frequency) is one logic descriptor, but this type of information is so important that it is dealt with separately under 'Spectral Analysis' (Chapter 7). The energy E could not be used as a single descriptor of 'size', as for this case

$$E \to \infty \quad \text{for} \quad t \to \infty$$

The rate of energy per period would, however, be constant, hence we use the power

$$P = \frac{1}{T} \int_0^T x^2(t) dt \quad \text{for} \quad T \to \infty$$

$$P \to \frac{1}{T_t} \int_0^{T_t} x^2(t) dt \tag{2.2}$$

as a descriptor, and such a signal, having finite power, is called a 'power signal'. As per the prior discussion, its units would be of V^2, N^2, etc.

2.2.3 *Random Signals (Cooper and McGrillen, 1971; Bendat and Piersol, 2000)*

Only statistical properties can describe random signals. The most basic statistical descriptions are those of probability distributions. Amplitude probabilities are defined for time signals, based on the

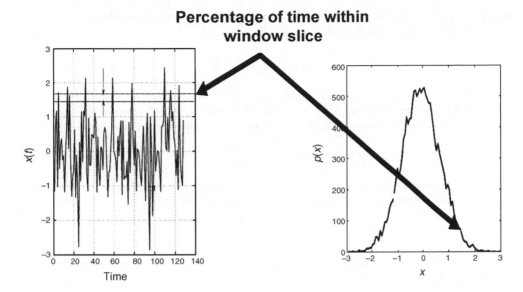

percentage of time for which a signal lies in a specific amplitude range. A probability density function $p(x)$ (continuous) is defined as (Figure T2.2):

$$p(x) = \lim_{\Delta x \to 0} \frac{\mathrm{Prob}[x < x(t) < x + \Delta x]}{\Delta x} = \lim_{\Delta x \to 0} \frac{1}{\Delta x} \left[\lim_{T \to \infty} \frac{\sum T_i}{T} \right] \qquad (2.3)$$

where T_i are the intervals where the signal lies in the amplitude window between x and $x + \Delta x$ (Figure T2.2).

The area under $p(x)$ is set to 1 for normalization. The area under $p(x)$ for a range of x is then

$$\int_{x_1}^{x_2} p(x) \mathrm{d}x \qquad (2.4)$$

being the percentage of time where the signal is between x_1 and x_2.

Signal parameters can be based on $p(x)$. In practice, statistical parameters will be computed over time, using consecutive signal samples. In signal processing, it is of advantage to process after removing the DC (mean value) of signals. Central moments, around the mean, are then defined as

First moment (mean): $\qquad \mu_1 = E[(x - \bar{x})] = \int_{-\infty}^{\infty} (x - \bar{x}) p(x) \mathrm{d}x$

Second moment (variance): $\qquad \mu_2 = E[(x - \bar{x})^2] = \int_{-\infty}^{\infty} (x - \bar{x})^2 p(x) \mathrm{d}x \qquad (2.5)$

Using the second one, variance, its square root will be the standard deviation.

For signals, statistical averages are computed as time averages. Then

$$\mu_x = \lim \frac{1}{T} \int_0^T x(t) \mathrm{d}t \quad T \to \infty \qquad (2.6)$$

assuming $\mu_1 = 0$:

$$\mathrm{Var}_x = \sigma_x^2 = \lim \frac{1}{T} \int_0^T x^2 \mathrm{d}t \quad T \to \infty$$

Hence Var$_x$ is also called the mean square (MS), and σ the root mean square (RMS).
For a random signal with zero mean,

$$x_{RMS} = \left[\frac{1}{T}\int_T x^2(t)dt\right]^{0.5} \tag{2.7}$$

The RMS can computed for any power signal, including deterministic ones. For analytical periodic functions it is possible to find the RMS as a function of amplitude. Many random phenomena have distributions which approximate the Gaussian distribution, also called the Normal distribution:

$$p(x) = \frac{1}{\sigma\sqrt{2\pi}}\exp\left[-\frac{(x-\mu)^2}{2\sigma^2}\right] = N(\mu,\sigma) \tag{2.8a}$$

$p(x)$ is described by two parameters only, the mean μ and the variance σ^2. The spread (width) of this bell-shaped function depends on σ. A normalized function is defined by

$$N(0,1) = p(x) = \frac{1}{\sqrt{2\pi}}\exp\left(-\frac{x^2}{2}\right) \tag{2.8b}$$

For 99% of the time a Gaussian signal is practically within $\pm 3\sigma$, which can thus be taken as an approximation of the peak values:

$$X_{\text{peak-peak}} = \pm 3\sigma = \pm 3x_{RMS}$$

2.3 Correlation Functions (Bendat and Piersol, 2000)

The notion of correlation is one of the most basic ones in the description of data, and especially joint descriptions. Joint descriptions between data points, whether from single or joint data sets, can describe patterns existing in the data.

Basic concepts for assessing the degree of linear dependence between two data sets x_k and y_k are based on covariance and correlation coefficients. The correlation coefficient R is then

$$R = \frac{\sum\limits_k x_k y_k}{\left(\sum\limits_k x_k^2\right)^{1/2}\left(\sum\limits_k y_k^2\right)^{1/2}} \tag{2.9}$$

The notion of these descriptors of dependence can be extended to time history data. Thus we have the covariance function between time data $x(t)$ and $y(t)$ as

$$C_{xy}(\tau) = E\left[\{x(t)-\mu_x\}\{y(t+\tau)-\mu_y\}\right] \tag{2.10}$$

where $E[.]$ denotes expectations, and $\mu_x = E[\{x(t)\}$ $\mu_y = E[\{y(t)\}]$.
For zero mean functions, we may use the cross-correlation function

$$R_{xy}(\tau) = E[x(t)y(t+\tau)] \tag{2.11}$$

which for continuous functions would be computed as

$$R_{xy}(\tau) = \lim\frac{1}{T}\int_0^T x(t)y(t+\tau)dt \quad T\to\infty \tag{2.12}$$

For the special case where $x(t) = y(t)$ we have the autocovariance and autocorrelation functions $C_{xx}(\tau)$, $R_{xx}(\tau)$. These functions depend on the specific dynamic pattern of the signals at hand, as they basically describe the correlation between signal points separated by τ.

Some general properties of the autocorrelation function are:

(a) The autocorrelation is an even function of τ

$$R_{xx}(\tau) = R_{xx}(-\tau) \qquad (2.13a)$$

(b) For $\tau = 0$

$$R_{xx}(0) = \sigma_x^2 + \mu_x^2 \qquad (2.13b)$$

where $\sigma_x^2 = E[(x - \mu_x)^2]$ is the variance of $x(t)$. For zero mean data (the prevalent case for vibration signals), $R_{xx}(0)$ is the mean square and thus also the power (the square of the RMS) of the signal.

The cross-correlation function has the inequality

$$\left| R_{xy}(\tau) \right| \leq \left[R_{xx}(0) R_{yy}(0) \right]^{1/2} \qquad (2.14)$$

2.4 Estimation and Errors (Bendat and Piersol, 2000)

2.4.1 The Notion of Realization

A single acquired random signal record can be considered as a sample, or realization, from a random population. There exists a variability between the various realizations from the same random process, even for stationary random signals. A variability is inherent in any parameter (mean, MS, RMS) and analysis-related functions (probability distribution, spectra, correlation–all to be introduced later). This variability is a fundamental property of any analysis method, implying the need to understand and hopefully control the uncertainty inherent in the results. One result of this variability is that any signal parameter or describing function can only be estimated. Knowledge concerning the error between the estimated and true parameter (or describing function) is of the greatest importance. Such errors are also called statistical sampling errors.

2.4.2 Denoting the Estimated Value of a Parameter p by $\hat{p}$

We define the two basic errors which are usually addressed:

Bias error: $\qquad\qquad\qquad\qquad e_b = E[\hat{p}] - p \qquad (2.15a)$

where $E[]$ denotes expectation and is estimated as mean

$$E[\hat{p}] = \frac{1}{N} \sum_1^N p_i$$

with p_i as the individual parameter computed from the ith observation. The expectation is defined as the limit for $N \to \infty$, while a finite N results in the mean p.

Random error: $\qquad\qquad e_r = \left\{ \lim_{N \to \infty} \left[\frac{1}{N} \sum_1^N (\hat{p}_i - E[\hat{p}])^2 \right] \right\}^{1/2} \qquad (2.15b)$

and the normalized errors e_b/p and e_r/p are used in practice.

The random error can often be reduced by increasing N, the number of data points used in the estimation of the parameter. However, this is not always true, as will be encountered when we estimate the power spectral density (Chapter 7).

3

Fourier Methods

Many signal processing tasks are performed in the frequency domain. The major tools used for this are based on Fourier methods, presented in this chapter (see Ambardar, 2007; Haykin and Van Veen, 1999; Gabel and Roberts, 1987).

It is convenient to describe various tools using Fourier methods according to some classification of methods:

- Methods for continuous signals:
  - Fourier series
  - Fourier transform
- Methods for discrete signals:
  - The discrete fourier transform.

While only the last is used in digital signal processing, the above order of presentation was deemed preferable from a didactic point of view.

3.1 Fourier Series

Fourier series are a special case of presenting or approximating functions by a set of mutually orthogonal functions. Specifically these series are used to decompose periodic signals into harmonic components.

Assuming a periodic function

$$x(t) = x(t + T)$$

we can decompose it as a Fourier series (FS). Various presentation of the FS are in use.

3.1.1 The Trigonometric Form

$$x(t) = \frac{a_0}{2} + \sum_{k=1}^{\infty} [a_k \cos(2\pi k f_0 t) + b_k \sin(2\pi k f_0 t)] \quad \text{with } f_0 = 1/T \tag{3.1a}$$

Discover Signal Processing: An Interactive Guide for Engineers S. Braun
© 2008 John Wiley & Sons, Ltd

which can also be presented in a polar form

$$x(t) = \sum_{k=0}^{\infty} C_k \cos(2\pi k f_0 t + \theta_k)$$

with   (3.1b)

$$C_k = [a_k^2 + b_k^2]^{0.5} \qquad \theta_k = \arctan\left(\frac{b_k}{a_k}\right)$$

The coefficients a_k and b_k are computed as

$$a_k = \frac{2}{T} \int_T x(t)\cos(2\pi k f_0)dt$$ (3.2)

$$b_k = \frac{2}{T} \int_T x(t)\sin(2\pi k f_0 t)dt$$

For $k = 1$ we have the fundamental components, the ones for $k > 1$ are the higher harmonics. The mean (or DC), of zero frequency, is represented by the term a_0.

3.1.2 The Exponential Form

$$x(t) = \sum_{k=-\infty}^{\infty} X_k \exp(j2\pi k f_0)$$

with   (3.3)

$$X_k = \frac{1}{T} \int_T x(t)\exp(-j2\pi k f_0 t)dt$$

X is complex, hence dictating the phase of the harmonic components.

The trigonometric form shows a one-sided presentation, with positive frequencies only. The exponential form, however, shows a two-sided presentation, which includes positive as well as negative frequencies. This is a purely mathematical convenience, resulting in more compact presentations. The negative frequencies obviously have no physical meaning, but have to be considered when total power is computed. The engineering units (EU) of X_k are those of $x(t)$ – volt, G, newton, etc.

As an example, applying Equations (3.1) and 3(2) to a unit amplitude, symmetric zero mean square signal, results in

$$x(t) = \frac{4}{\pi}\left[\sin(\omega_0 t) + \frac{1}{3}\sin(3\omega_0 t) + \frac{1}{5}\sin(5\omega_0 t)...\right] \qquad \omega_0 = \frac{2\pi}{T}$$

A 2D graphical representation can represent the spectrum. For the trigonometric form, the one-sided presentation, this has the form of lines at the corresponding frequencies (Figure T3.1a). For the complex presentation with two-sided spectra, all magnitudes (except the DC at frequency zero) have a value of 50% as compared to the one (Figure T3.1b)

For a specific time signal, a sinusoid, with a frequency of f_p, spanning exactly p periods, an interesting observation is possible. Let us assume a mathematical Fourier series decomposition based on a basic period equaling the spanned time T_{total}, and hence a fundamental frequency of $1/T_{total} = f_{fundamental}$. The signal's basic physical period is $T_p = T_{total}/p$, and its frequency $f_p = p/T_{total}$. Hence the spectral line describing this signal will occur at a frequency of

$$k f_{fundamental} = f_p$$

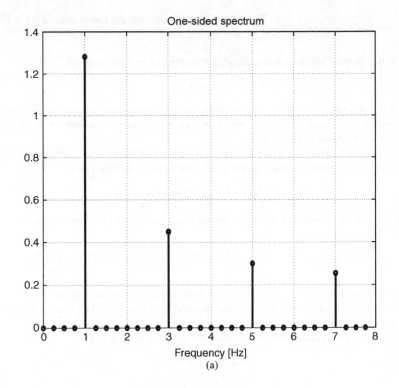

One-sided spectrum

(a)

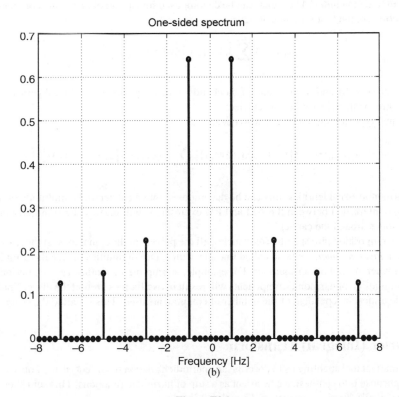

One-sided spectrum

(b)

Figure T3.1

The index of the spectral line will thus indicate the number of periods spanned

$$k = p$$

In the spectral representation the line corresponding to this sinusoid hence occurs at the pth harmonic.

Other examples of periodic signal are modulated ones. We will encounter these when analyzing bearing and gear vibrations (Chapter 13).

A carrier signal of frequency f_c is modulated by a signal with frequency f_m. These are obviously of periodic character. For amplitude modulation (AM), the signal is modeled as

$$x_{AM}(t) = A[1 + k_{AM}\cos(2\pi f_m t)]\cos(2\pi f_c t)$$

and using trigonometric identities this can be decomposed into a sum of harmonic signals

$$x_{AM}(t) = A\cos(2\pi f_c t) + \frac{Ak_{AM}}{2}\{\cos[2\pi(f_c + f_m)t] + \cos[2\pi(f_c - f_m)t]\}$$

with the three frequencies f_c, $f_c + f_m$, $f_c - f_m$. The first component is the carrier, the two others the sidebands, whose amplitude is a function of the modulation index k_{AM}. The sideband amplitudes are symmetric around the carrier.

For frequency (FM) or phase (PM) modulation, the signal is modeled as

$$X_{PM}(t) = A\cos(O)$$

$$O = 2*\pi*f_c t + k_{PM}\cos(2\pi f_m t)$$

with k_{PM} a modulation index. The signal can be decomposed into a sum of harmonic components using Bessel functions of the first kind, resulting in

$$x_{PM} = A\sum_{k=-\infty}^{\infty} J(k_{PM})\cos[2\pi(f_c + kf_m)t]$$

with a carrier ($k = 0$) and a multitude of sidebands with frequencies $f_c \pm kf_m$. Again the sideband amplitudes are symmetric around the carrier.

A combined type of modulation can also occur:

$$x_{AM-PM}(t) = A[1 + k_{AM}\cos(2\pi f_m t)]\sum_{k=-\infty}^{\infty} J(k_{PM})\cos[2\pi(f_c + kf_m)t]$$

Applying trigonometric identities, this can be decomposed into a carrier and a multitude of sidebands. However, the interaction between the AM and PM expressions will destroy the symmetry of the sideband amplitudes around the carrier.

With a decomposition into specific frequencies, all the powers of the signal exist at these frequencies, resulting in a *discrete spectrum* (also called line spectrum). Discontinuities in signals result in specific phenomena when using Fourier methods. For example, attempting to synthesize such periodic signals by finite summations of harmonic components will result in oscillations around the discontinuities, with a magnitude almost independent of the number of components used. This is called the **Gibbs** effect.

3.2 Fourier (Integral) Transform

The mathematical tool enabling us to describe an aperiodic function (transient) is the Fourier transform (FT). The purpose is to synthesize a transient as a sum of harmonic functions. This sum should be zero outside the signal's duration interval, and equal to it inside it.

The definition of this transform is

$$x(t) = \int_{-\infty}^{\infty} X(f) \exp(j2\pi f) df$$

$$X(f) = \int_{-\infty}^{\infty} x(t) \exp(-j2\pi ft) dt$$

(3.4)

or symbolically

$$x(t) \leftrightarrow X(f)$$

$x(t)$ and $X(f)$ are a Fourier transform pair, with x in the time and X in the frequency domain.

A range of continuous frequencies is needed to represent $x(t)$ by a sum of harmonic functions, basically in a two-sided presentation. An aperiodic function is characterized by a *continuous spectrum*.

The EU of the $X(f)$ (based on $X(f) = TX_k$) are then V-sec, G-sec, newton-sec, etc. The Fourier representation for random signals will be addressed in Chapter 7 (Spectral Analysis). The EU for these signals will be seen to be [V^2/Hz].

As with general transforms (Laplace, Z), specific transforms and general properties of the transformation can be very useful. Denoting by $F[.]$ a Fourier transform, the following theorems are noted:

$$F[a_1 x_1 + a_2 x_2] = a_1 F(x_1) + A_2 F(x_2)$$

$$F(at) = \frac{1}{|a|} X\left(\frac{f}{a}\right)$$

(3.5)

$$F[\exp(j2\pi f_0 t)] = \delta(f - f_0)$$

$$F[x(t) \exp(2\pi j\alpha t)] = X(f - \alpha)$$

The first one is the theorem of linearity. The second one is the scaling theorem, with the constant a being a scalar. Compression and dilation of $x(at)$ is achieved by $a > 1$ and $a < 1$ respectively. The scaling theorem indicates that compressing a signal will widen its spectrum, and vice versa. The division by abs(a) is intended to keep the energy in the time and frequency domain invariant.

The third theorem is helpful in finding the Fourier transform (as opposed to the Fourier series representation) of a periodic signal. Applying it to the FS of $x(t)$,

$$x(t) = \sum_{k=-\infty}^{\infty} X_k \exp(j2\pi k f_0 t) \leftrightarrow \sum_{k=-\infty}^{\infty} X_k \delta(f - k f_0)$$

(3.6)

Theoretically, impulses result only when the limits of integration (Equation 3.4) tend to infinity. The FT then has the form of impulses located at $k f_0$, the frequencies of the Fourier series components.

The fourth one is the shifting theorem, to be used in Chapter 10.

3.3 The Uncertainty Principle

This fundamental principle relates the frequency bandwidth occupied by a signal B to its duration T in time. These are related by

$$BT \geq k$$

(3.7a)

where k depends on how B and T are specifically defined. k is usually in the range of 0.5–2.

One important limitation can be directly traced to this principle – the possibility to identify in a compound signal two signal components of frequencies f_1 and f_2 by any analysis. Unless a minimum time duration T is available for analysis, such that

$$T \geq k(|f_1 - f_2|)$$

the separation is impossible. There is always an inverse relationship between the signal duration and the minimal detectable frequency separation between any of its components.

The effect of truncating a signal can be understood intuitively from this principle: The time available ΔT around the truncation phase tends to zero, resulting in a spread (i.e. high frequencies) in the frequency domain.

3.4 The Discrete Fourier Transform (DFT)

This is formally called the discrete time fourier series (DTFS), but the term DFT has been accepted in practice. It shows a relation between two periodic sequences, $x(n)$ in time and $X(k)$ in the frequency domain.

$$X(k) = \sum_n x(n) \exp\left(-j\frac{2\pi}{N} kn\right)$$

$$x(n) = \frac{1}{N} \sum_k X(k) \exp\left(j\frac{2\pi}{N} kn\right)$$

(3.8a)

The normalization factors (1 for $X(k)$ and $1/N$ for $x(n)$) are conventions. Others exist, for example $1/\sqrt{N}$ in both expressions. It is always recommended to check which one is used for specific computations, as this affects normalization factors appearing in expressions involving the DFT. Equation (3.8a) is the one prevailing in engineering. Both sequences are periodic in N. $x(n)$ and $X(k)$ are a DTFS pair, denoted as

$$x(n) \leftrightarrow X(k)$$

N values of $X(k)$ can be obtained from N values of $x(n)$ and vice versa. Extending n or k beyond N will then generate the periodic discrete sequences.

Due to the periodicity in N, the DFT relations are summarized as

$$X(k) = \sum_{n=0}^{N-1} x(n) \exp\left(-j\frac{2\pi}{N} kn\right)$$

$$x(n) = \frac{1}{N} \sum_{k=0}^{N-1} X(k) \exp\left(j\frac{2\pi}{N} kn\right)$$

(3.8b)

with a frequency spacing of $1/N$ for $X(k)$. The range of k can easily be modified so as to include negative values, one possibility being that of 0 to $N/2$ and 0 to $N/2-1$. Thus negative indices k correspond to negative frequencies $-k\Delta f$, with

$$\Delta f = 1/N$$

(3.9a)

While a sequence with a normalized interval of $\Delta t = 1$ is often assumed in this chapter, the real time interval Δt will be considered again in later chapters (for example Chapter 7 on spectral analysis). The frequency spacing for the computed values will then be

$$\Delta f = \frac{1}{N\Delta t}$$

(3.9b)

and this is the computable frequency spacing dictated by the analysis parameters. For the case of real signals $x(n)$, specific symmetries must exist in $X(k)$, as $x(n)$, the inverse DFT of $X(k)$, has zero imaginary values (see Figure T3.2). The symmetries are

$$X(-k) = X(k)^*$$

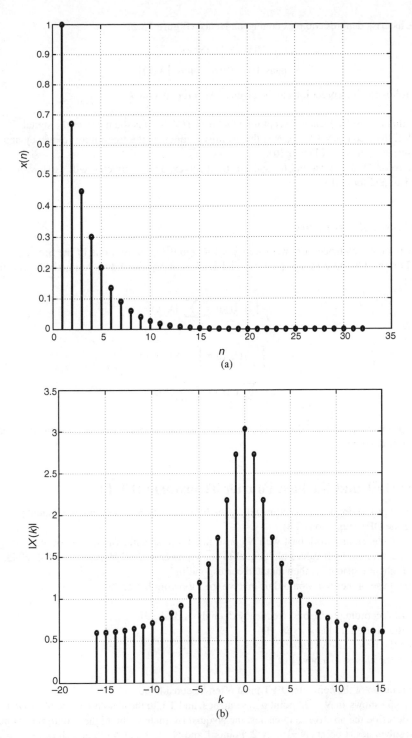

(a)

(b)

Figure T3.2

where the asterisk denotes complex conjugate. Hence (Figure T3.2)

$$|X(-k)| = |X(k)|$$
$$\text{phase}[X(-k)] = -\text{phase}[X(k)]$$

(3.10)

Thus the following interpretation can be given to $X(k)$ for specific k:

1. $k = 0$, this is zero frequency. Except for a factor of $1/N$, X(0) is the mean of the signal.
2. $k = N/2$, the frequency $k\Delta f = k/N$ is the Nyquist frequency, the maximum frequency appearing in the discrete signal (see Chapter 10).
3. $k = -1$ to $-N/2-1$ correspond to negative frequencies. The symmetry will exist with $X(k)$ in the range 1 to $N/2$ for real x.

3.4.1 Parseval Relations

There exists an equivalence in power or energy for signal representations in the time and frequency domain. For the FS (power), FT (energy) and DFT (energy) we have, respectively, the following equalities:

$$\frac{1}{T}\int_T x^2(t)dt = \sum_{k=-\infty}^{\infty} |X(k)|^2$$

$$\int_{-\infty}^{\infty} x(t)^2 dt = \int_{-\infty}^{\infty} |X(f)|^2 df$$

(3.11)

$$\sum_{n=0}^{N-1} x^2(n) = \frac{1}{N}\sum_{k=0}^{N-1} |X(k)|^2$$

The cross products $X(k_1)^*X(k_2)$ do not add to the total power (energy) due to the orthogonal properties of the Fourier decomposition.

3.5 The DFT and the Fast Fourier Transform (FFT)

Digital signal processing was revolutionized by the FFT algorithm, an extremely efficient procedure for computing the DFT (Equations 3.8).

Many FFT algorithms work best with N constrained to a specific value. One of the most prevalent is the radix 2 version, with values of N equal to some power of 2, $N = 16, 32,...1024, 2048$, etc. The resulting frequency spacing is then dictated correspondingly.

A practical use of the FFT would thus involve the following hierarchy:

- Choose Δt (according to highest frequency of analysis desired)
- Choose the desired frequency spacing
- Compute the necessary N (Equation 3.9b)
- Choose N the closest to a power of 2.

The interpretation of the computed FFT must often be considered.

Figure T3.3a shows an $N = 32$ point real sequence, and T3.3b the absolute value of its DFT. Points 0 to $N/2-1$ describe the positive frequencies, the Nyquist frequency (the highest analyzed) at point $N/2$, negative frequencies at points $N-1$ to $N/2$. Points 1 and $N-1$, 2 and $N-2$, etc. show the symmetry of complex conjugate values (absolute values are obviously equal). A shifted version is needed to correspond to the accepted symmetric positive–negative frequency display (Figure T3.3c).

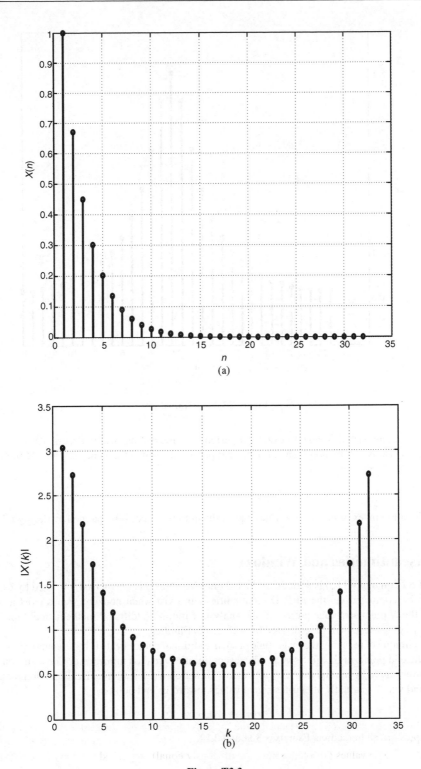

Figure T3.3

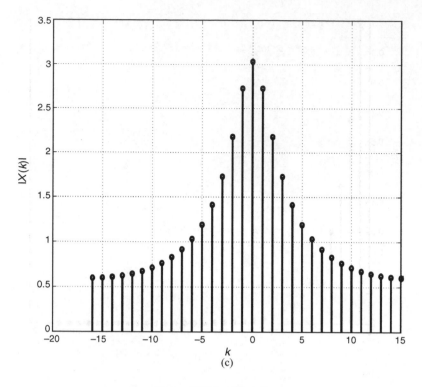

(c)

Figure T3.3 (*continued*)

For a harmonic signal, spanning exactly p periods, an interesting results occurs. Assume a signal spanning p periods, N total points and M points per period: the signal frequency $k\Delta f = 1/M\Delta t$, $M = N/p$. With $\Delta f = 1/N\Delta t$, this results in

$$k = p \tag{3.12}$$

and the location index of the spectral lines equals the number of periods spanned by the signal.

3.6 Discontinuities and Windows

The DTFS (Equations 3.8a) is periodic in N, and only N values (i.e. one period) are used by Equations (3.8b) to be computed via the FFT. Discontinuities can exist when nonzero values exist at the end point of the N point sequence present in the analysis window. Additional mathematical, nonphysical frequency components will be generated by such discontinuities.

One technique of reducing the magnitude of such additional components is by attenuating the magnitude of the end points of the time signal. The discontinuities are then minimized. The technique modifies the time signal by multiplying it by a 'window' function $w(n)$, a symmetrical function maximized at $N/2$, and which gradually attenuates signal values toward its end points:

$$x(n) \rightarrow w(n)x(n) \tag{3.13}$$

These topics are addressed via Exercises 3.11 and 3.12.

4

Linear Systems

This chapter deals only with linear time invariant systems (see Gabel and Roberts, 1987; Haykin and Van Veen, 1999). Both continuous and discrete ones are described.

4.1 Continuous Systems

The most basic description is that of an ordinary differential equation:

$$\sum_{k=0}^{N} a_k \frac{d^k}{dt^k} y(t) = \sum_{k=0}^{M} b_k \frac{d^k}{dt^k} x(t) \tag{4.1a}$$

the system being defined by the equation orders and parameters a_k and b_k.

Algebraic relations can be obtained via transform methods. For the Laplace transform

$$\sum_{k=0}^{N} a_k s^k Y(s) = \sum_{k=0}^{M} b_k s^k X(s) \tag{4.1b}$$

via which a transfer function can be obtained as

$$H(s) = \frac{Y(s)}{X(s)} = \frac{\sum_{k=0}^{M} b_k s^k}{\sum_{k=0}^{N} a_k s^k} = \frac{B(s)}{A(s)} \tag{4.2}$$

The zeros of the rational polynomials in s are another description:

$$A(s) = \prod_{k=1}^{N} (s - z_k)$$

$$B(s) = b_m \prod_{k=1}^{M} (s - p_k) \tag{4.3}$$

$$H(s) = \frac{B(s)}{A(s)}$$

with z_k and p_k the zeros and poles of the system.

The FRF (frequency response function) is

$$H(j\omega) = H(s)\big|_{s=j\omega} \tag{4.4a}$$

Discover Signal Processing: An Interactive Guide for Engineers   S. Braun
© 2008 John Wiley & Sons, Ltd

i.e by evaluating H on the $s = j\omega$ contour. The FRF is one of the most widely used descriptors, as it has a very intuitive interpretation. Assuming a harmonic excitation

$$x(t) = X_{max} \sin(\omega t)$$

the steady state response of any linear system will be of the form

$$y(t) = Y_{max} \sin(\omega t - \phi)$$

and the FRF is then a vector whose modul is the gain Y_{max}/X_{max} (a function of frequency) with an angle corresponding to the phase shift ϕ (also a function of frequency):

$$H(\omega) = |H(\omega)| \angle \phi(\omega) \tag{4.4b}$$

The two presentations $|H(\omega)|$ and $\angle \phi(\omega)$ are known as Bode plots, showing the gain and phase shift as a function of frequency. The system can also be described via its impulse response $h(t)$, the result of applying an excitation of the form

$$x(t) = \delta(t)$$

resulting in $h(t)$. In can be shown that $h(t)$ is the inverse Laplace transform of $H(s)$ or that of the Fourier transform $H(\omega)$. Thus we have the basic Fourier transform relationship

$$h(t) \leftrightarrow H(\omega) \tag{4.5}$$

An important problem in the area of systems is the computation of the response $y(t)$ to an excitation $x(t)$. For time-invariant linear continuous systems we may use the impulse response via the convolution operation

$$y(t) = \int_{-\infty}^{\infty} x(\tau)h(t-\tau)d\tau = x(t) \otimes h(t) \tag{4.6a}$$

or applying a Fourier transform (FT) to both sides (the FT of a convolution is the products of the FTs)

$$Y(\omega) = H(\omega) X(\omega) \tag{4.6b}$$

This last expression is intuitively obvious, at each frequency the response being the (complex) product of the input and the gain.

4.2 Discrete Systems

The most basic description is that of an ordinary difference equation:

$$\sum_{k=0}^{N} a_k y(n-k) = \sum_{k=0}^{M} b_k x(n-k) \tag{4.7a}$$

the system being defined by the equation orders and parameters a_k and b_k.

Algebraic relations can be obtained via transform methods. For the Z transform

$$\sum_{k=0}^{N} a_k z^{-k} Y(z) = \sum_{k=0}^{M} b_k z^{-k} X(z) \tag{4.7b}$$

via which a transfer function can be obtained as

$$H(z) = \frac{B(z)}{A(z)} = \frac{\sum\limits_{k=0}^{M} b_k z^{-k}}{\sum\limits_{k=0}^{N} a_k z^{-l}} \tag{4.8}$$

The zeros of the rational polynomials in z are another description:

$$H(z) = b_m \frac{\prod\limits_{k=j}^{M}(z - z_k)}{\prod\limits_{k=1}^{N}(z - p_k)} \qquad (4.9)$$

with z_k the zeros and p_k the poles of the system.

For the FRF, we will assume samples separated by a time interval Δt. The frequency response of a discrete system is given by

$$H(j\phi) = H(z)\Big|_{z=j\phi}$$

$$\phi = \omega\Delta t \quad H(j\omega) = H(z)\Big|_{z=\exp(j\omega\Delta t)} \qquad (4.10a)$$

i.e. evaluating $H(z)$ on the $z = \exp(j\omega t)$ contour, with the sampling interval dt. Note that this FRF is a *continuous* function of the frequency. It is also periodic with $2\pi/\Delta t$.

In a similar fashion to that already mentioned for continuous systems, the FRF is also one of the most widely used descriptors for discrete ones. Again assuming a harmonic excitation

$$x(n\Delta t) = X_{max} \sin(\omega n\Delta t)$$

The steady state response of any linear system will be of the form

$$y(n\Delta t) = Y_{max}\sin(\omega n\Delta t - \phi)$$

In this context, harmonic digital signals are sequences corresponding to the sampled values of continuous harmonic signals. One convenient point of view is top 'look' at the sequence as having a fictional harmonic envelope.

The FRF is then a vector whose modul is the gain Y_{max}/X_{max} (a function of frequency) and phase shift ϕ (also a function of frequency):

$$H(\omega) = |H(\omega)|\angle\phi(\omega) \qquad (4.10b)$$

The two presentations $|H(\omega)|$ and $\angle\phi(\omega)$ are known as Bode plots, showing the gain and phase shift as a function of frequency.

The system can also be described via its impulse response $h(n)$, the result of applying an excitation of the form

$$x(n) = \delta(n)$$

resulting in $h(n)$. It can be shown that $h(n)$ is the inverse Z transform of $H(z)$. Thus we have the basic transform relationship

$$h(n) \leftrightarrow H(z) \qquad (4.11)$$

As before, we could use $h(n)$ to compute the response $y(n)$ to an excitation $x(n)$. For time-invariant linear discrete systems we use the impulse response via the discrete convolution operation

$$y(n) = \sum_{k=-\infty}^{\infty} x(k)h(n-k) = h(n) \otimes x(n) \qquad (4.12a)$$

or applying the Z transform to both sides

$$Y(z) = H(z)X(z) \qquad (4.12b)$$

It is often desired to design discrete systems whose characteristics approximate those of specific continuous systems. The degree of approximation can often be dictated (at the expense of the system's complexity), but an ideal equivalence is not achievable in practice. Various methods are available for these approximations, some of them used in Chapter 5 (filters).

4.3 A Specific Case of a Continuous Linear Systems – Accelerometers (Dally *et al.*, 1993; Doebelin, 2003)

Accelerometers are inertial based vibration sensors: a mass spring system is base excited by the displacement of the body to which it is attached. An appropriate transducing device measures the force sensed by the spring. Some sensing devices are based on strain gauges. One of the most popular accelerometer types is the piezoelectric one, where the force sensed by the spring is sensed by a piezoelectric transducer generating a charge (which can often generate a proportional voltage). The dynamic characteristics of such an accelerometer are thus dictated by both the mechanical and piezoelectric subsystems (Figure T4.1).

The mechanical transfer function relating the response (spring deflection) to the excitation (acceleration) is

$$H(s) = \frac{1}{\omega_0^2} \frac{1}{\dfrac{s^2}{\omega_0^2} + 2\zeta \dfrac{s}{\omega_0} + 1} \tag{4.13}$$

where ω_0 is the undamped natural frequency and ζ the damping ratio. Thus the accelerometer behaves like a second-order system, operated below resonance. In the frequency range below ω_0, and very small damping ratios, the sensed force $x_f(t)$, resulting from a base acceleration $\ddot{x}(t)$, is then

$$x_f(t) = \frac{k}{\omega_0^2} \ddot{x}(t) \tag{4.14}$$

with k the spring constant (newton/m), and for equal design the gain is inversely proportional to the square of the natural frequency.

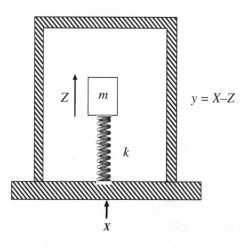

Figure T4.1

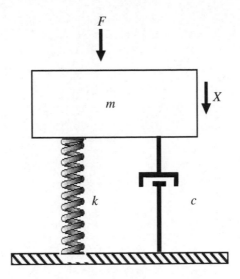

Figure T4.2

Appendix 4.A The Lightly Damped SDOF System

Second-order low damped SDOF system

As an example, we analyze the mass–spring–dashpot system (Figure T4.2):

$$m\frac{d^2x}{dt^2} + c\frac{dx}{dt} + kx = F$$

$$(-\omega^2 m + j\omega c + k)X(\omega) = F(\omega)$$

$$X(\omega) = \frac{1}{-\omega^2 m + j\omega c + k}F(\omega) = H(\omega)F(\omega)$$

$$\Omega = \frac{\omega}{\omega_0}$$

$$\omega_0 = \left[\frac{k}{m}\right]^{0.5} \quad \varsigma = \frac{c}{2\left[mk\right]^{0.5}}$$

with ω_0 the undamped natural frequency and ς the damping ratio,

$$H(\Omega) = \frac{1/k}{1 - \Omega^2 + 2j\varsigma\Omega}$$

The absolute FRF is approximately like that shown by Figure T1.3. The frequency at which $|H|$ is maximum is known as the resonance frequency:

$$\omega_r = \omega_0\left[1 - 2\varsigma^2\right]^{0.5} \approx \omega_0$$

For $\varsigma \ll 1$ the 3 dB bandwidth BW can be shown to be

$$\text{BW} = 2\varsigma\omega_0 \text{ in rad/sec} \quad \text{or} \quad 2\varsigma\frac{\omega_0}{2\pi} = 2\varsigma f_0 \text{ in Hz}$$

5

Filters

5.1 Preliminaries

Filtering is a time domain signal processing operation, applying a predetermined weighing to different frequency regions (Ambardar, 2007, Hamming, 1989). Typical applications are the separation of signal components, which are concentrated in different frequency regions, and the improvement of signal to noise ratio (S/N) by the rejection of undesired components.

The dynamic properties of the physical systems through which signals propagate, including the measurement system, usually apply a frequency-dependent weighing. It is often convenient to consider these as filters operating on the information carrying signals being analyzed. Such filters are of the analog type. Vibration signals are often highly affected by the physical system through which they propagate. Thus while this book deals mainly with digital signal processing, some aspects of analog filtering, and not only of digital ones, are also considered.

Spectral analysis, described in detail in Chapter 7, is a frequency domain representation, without the ability to describe localization in time of signal properties. This local information is retained by filtering, at the price of looking at one specific frequency region at a time. The spectral analysis and filtering can be considered as complementary operations, a different way of looking at signals in two domains. The major types of filters are characterized by their band pass and rejection ranges, via their frequency response function (FRF). Figure T5.1 shows this for some idealized filters.

As an example, we take the signal depicted by Figure T5.2. The signal itself could be of any physical dimension like voltage, force, acceleration, pressure etc. The general nature is one of a (very) noisy transient. The filtering operation depicted attempts to extract the transient occurring at the earlier part of the signal, attenuating the faster fluctuations superimposed on it. The filter has a sluggish response, thus attenuating the noise. Two degrees of filtering are shown. While the noise is indeed attenuated, the transient itself is also modified: the amplitude seems reduced, and a delayed – smeared – version is extracted. There is a trade-off between the smoothing of the transient (attenuation of the noise) and the modification: the higher the attenuation, the more severe the modification. This demonstrates one of the major problems of applying a filter: that of balancing the attenuation of undesired signal components and possible modifications of the desired one.

5.2 Analog and Digital Filters

Various realizations of filtering systems exist, depending on the type of physical system at hand. In the previous example, assuming a pressure signal, a pneumatic filter, consisting of a restriction and an attached volume, could have attenuated the noise. Thus we encounter mechanical filters consisting of spring masses and damping elements, electrical filters consisting of resistors, capacitors and inductances and so forth.

Discover Signal Processing: An Interactive Guide for Engineers   S. Braun
© 2008 John Wiley & Sons, Ltd

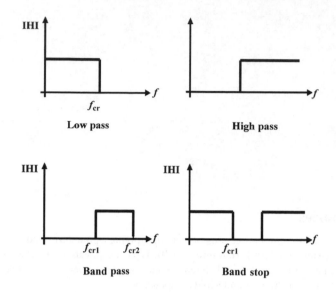

Figure T5.1

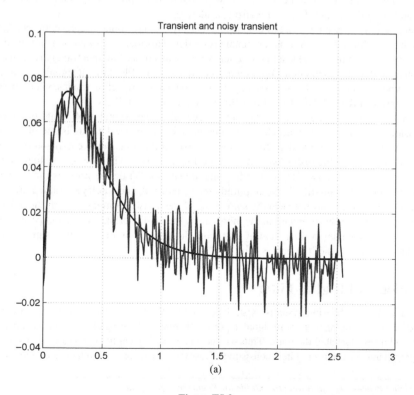

Figure T5.2

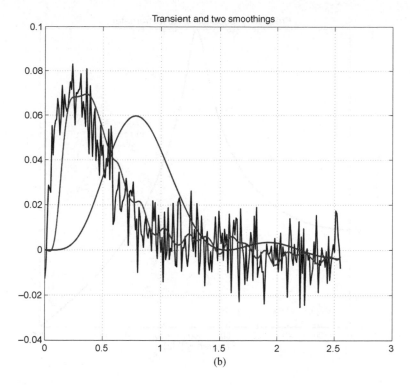

Figure T5.2 (*continued*)

Filtering effects often result from the monitored system and measuring hardware used. Looking at a complete measuring chain, the dynamic characteristics of the sensor (for example an accelerometer) constitute a specific type of analog filter.

As shown in Chapter 4, the FRF of a typical accelerometer, with a low damping factor, exhibits a large gain around the natural frequency of the sensor. (To aggravate matters, this natural frequency may be heavily dependent on the specific mounting method used.) For all practical purposes, the net effect is one of modifying, or filtering, the measured signal (Figure T5.3).

Filtering effects can be achieved by computations applied to digital signals, acquired via sensors and digitization systems. These algorithms are known as digital filters, operating on discrete time sequences. It is convenient to look at these algorithms as digital systems, designed so as to have the desired filtering properties (Figure T5.4).

The quantitative approach is obviously that of linear systems (Chapter 4). Systematic designs can be based on the frequency response function (FRF) and the impulse response (IR). The first deals with the case of harmonic excitation, giving the gain and phase shift as a function of frequency. The impulse response is the inverse Fourier transform of the FRF. The basic analysis starts with difference equations and the Z transform.

Digital filters fall into two general categories. The general discrete transfer function of a linear system is given by

$$H(z) = \frac{Y(z)}{X(z)} = \frac{\sum\limits_{r=0}^{r=m} b_r z^{-r}}{\sum\limits_{r=0}^{r=n} a_r z^{-r}} \qquad (5.1a)$$

which in the time domain would be, with $a_0 = 1$,

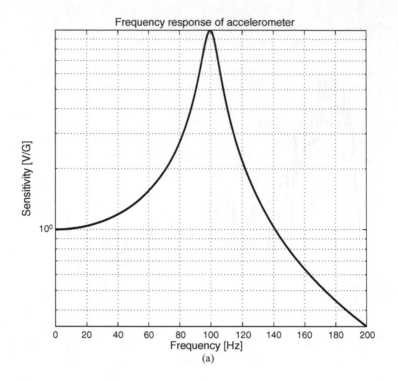

(a)

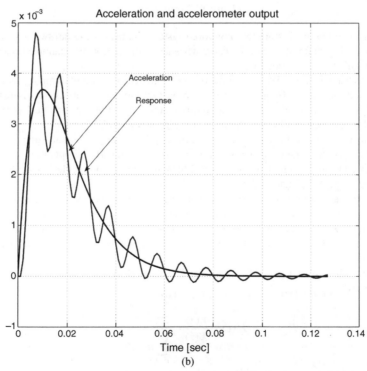

(b)

Figure T5.3

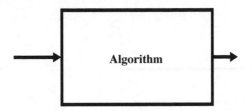

Figure T5.4

$$y(n) = -\sum_{r=1}^{r=n} a_r y(n-r) + \sum_{r=0}^{r=m} b_r x(n-r) \qquad (5.1\text{b})$$

For $m = 0$ we have an AR (autoregressive) or infinite impulse response (IIR) filter, whereas with $n = 0$ we have an MA (moving average) or finite impulse response (FIR) one.

$$H_{MA}(z) = b_0 + b_1 z^{-1} + b_2 z^{-2} \ldots + b_M z^{-M} = \sum_{r=0}^{M} b_r z^{-r} = \sum_{r=0}^{M} h_r z^{-r} \qquad (5.2)$$

the b_r being equal to the discrete impulse response elements h_r, whereas for the AR filter

$$H_{AR}(z) = \frac{1}{\displaystyle\sum_{r=0}^{r=n} a_r z^{-r}} = \sum_{r=0}^{\infty} c_r z^{-r} \qquad (5.3)$$

The IIR follows immediately from the polynomial division.

The two types of filters differ significantly in their properties:

(a) *Filter length*: Effective IIR filters will usually incorporate much fewer coefficients than the FIR one. A saving by a factor of 20 or more is quite common. The computational effort, and easier possibilities of real time computations, are an definite advantage of using IIR realizations.
(b) *Stability*: FIR filters are inherently stable, IIR may become unstable, and this aspect has to be taken into consideration when designing the filter.
(c) *Linear phase characteristics*: This is desirable in some applications (see later). One major advantage of FIR filters is the ease of achieving this property, even for real time uses. Unless special hardware solutions are used, the IIR filters can have this property only if used offline.

The filter's characteristics depend on the values of the parameters a_i and b_i in equation The design of a digital filter then consists in determining:

- The number of parameters a_i and b_i (the filter order)
- The numerical value of the parameters.

5.3 Filter Classification and Specifications

Filters can be classified according to their region of operations. Thus we have low pass, high pass, band pass and band stop types as shown in Figure T5.1. Shown were the responses of the ideal filters, with a discontinuity in the FRF. The ideal filters are not realizable, whether in analog or digital form, but they can be approximated. The approximation will result in nonideal filters. Deviations of the FRF from the ideal are often given as uncertainty bands (see Figure T5.5). Three types of possible nonidealized performance should be noted (Figure T5.5). First of all the gain in the pass band may fluctuate within some given

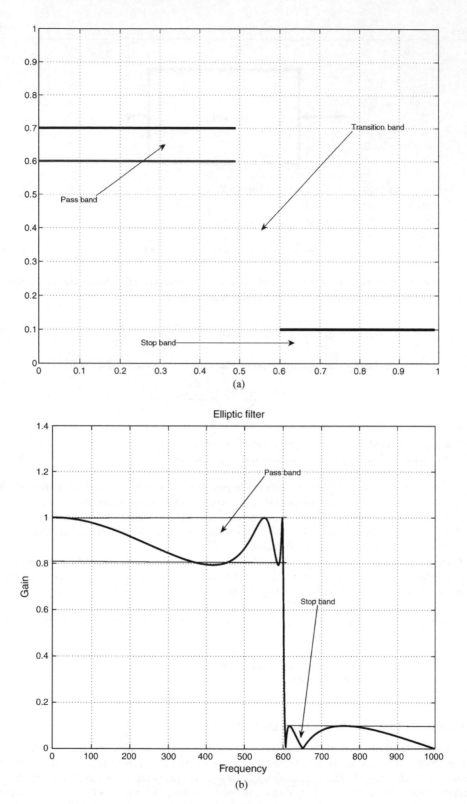

Figure T5.5

bounds. Next the attenuation in the stop band can be finite, and again fluctuate within other bounds. Finally the transition from pass to stop band is not discontinuous, but will exhibit a transition width.

Often a simplified specification will be used. The region of transition where attenuation exists is obviously the main specified parameter. These are given via 'critical' frequencies, one for the cases of low/high pass, two for the band pass/stop one. The transition may be gradual, and the critical frequencies are usually defined as the half power (or −3 dB) frequency points. The steepness is then the next parameter of importance. Many analog filter FRFs are asymptotic to straight lines in the log/log presentation. The steepness is then described by the slope of these asymptotes. Examples would be + 12 dB/octave for the high pass, or −18 dB/octave for the low pass case (an octave is a doubling of frequency, a term taken from music theory). For a large case of filters, the slope is in integer multiples of 6 dB/octave, i.e. 6, 12, 18 … 72, etc. This is related to the order of the differential equation describing the continuous (analog) filter. For digital filters, the order of the difference equation can dictate the steepness function, the shape of which is, however, often more complex.

One important practical observation can be made concerning the exact desired filter characteristics. In some disciplines, actual a priori information exists concerning the information-carrying signal. For example, in many communication tasks, the shape of signals is determined (and hence known) by the sender. This often enables us to prescribe optimum filter characteristics. In other disciplines, such exact information may not available. Exact filter requirements on the transition shape of filter FRFs may not be available. Thus often only the filter's type and basic characteristics (say critical frequencies and order) will be chosen.

One exception is the characteristic of the phase response. Linear phase filters have a phase characteristic of constant slope. They are often required for the measurement of transients.

From the user point of view, interaction with software would include two phases (Figure T5.6):

- the design of the filter (i.e. computation of coefficients a_n and b_n)
- the application of the filter (i.e solving the difference equations).

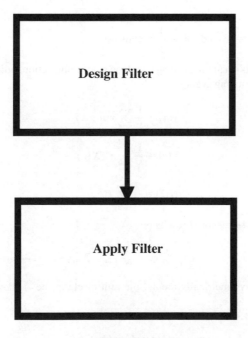

Figure T5.6

Depending on the available software, the design and filtering phases can sometimes be done via a graphical user interface (GUI).

5.4 IIR Filters

Some of the most popular design methods are based on prototype filters, which were originally developed for analog filters. The prototype-based design attempts to duplicate the performance of these historic analog filters. The main ones are:

- Butterworth filter
- Chebychev filter
- elliptic filter.

Butterworth filters have 'maximum flatness' properties, such that for an N order filter, $2N - 1$ derivatives of the FRF would be zero at zero frequency (for low pass filters).

Chebychev filters can have equal ripples in the bandpass (or bandstop) and maximum flatness in the bandstop (or bandpass). Thus the transition between the bands can be more rapid than for the Butterworth case. Elliptic filters can have equal ripples in the bandpass and bandstop. For a given order and ripple it can minimize the transition width.

Other methods can be used to approximate a given FRF, often based on optimization methods.

5.5 FIR Filters

We illustrate with a simple example:

$$y(n) = \frac{1}{3}x(n) + \frac{1}{3}x(n-1) + \frac{1}{3}x(n-2) = \sum_{k=0}^{N-1} b_k x(n-k)$$

with

$$N = 3 \quad \text{and} \quad b_k = \frac{1}{N} = \text{constant}$$

This is obviously a three point running average, and the smoothing action indicates a low pass filter. For the more general N point moving average

$$y(n) = \frac{1}{N}\sum_{k=0}^{N-1} x(n-k)$$

$$Y(z) = \frac{1}{N}\sum_{k=0}^{N-1} z^{-k} X(z) \tag{5.4a}$$

$$H(z) = \frac{1}{N}\sum_{k=0}^{N-1} z^{-k}$$

Summing the geometrical series for H results in

$$H(z) = \frac{1}{N}\frac{1-z^{-N}}{1-z^{-1}} \tag{5.4b}$$

with N zeros distributed symmetrically around the unit circle in the Z plane, and one pole at $z = 1$ (cancelling one zero).

Substituting Equation (4.10a) results in

$$H(z)\Big|_{z=\exp(j\omega\Delta t)} = \frac{1}{N}\frac{1-\exp(-iN\omega\Delta t)}{1-\exp(-i\omega\Delta t)} = \frac{1}{N}\frac{\sin(N\omega\Delta t/2)}{\sin(\omega\Delta t/2)}\exp\left[-j\omega(N-1)\Delta t/2\right] \tag{5.4c}$$

with $|H|$ characterized by a main lobe around zero frequency, which gets narrower as N increases – basically a sharp low pass filter.

As stated, one major feature of such filters is the ease of designing linear phase realizations. We now describe two popular design principles

5.5.1 The Window Method

This is based on specifying an FRF, and computing the corresponding impulse response by an inverse Fourier transform. As noted in Section 5.2, the coefficients b_r of the FIR filter are actually elements of the impulse response.

For finite bandwidth filters, the theoretical impulse response will be unlimited in time. This follows from a fundamental property of Fourier transforms, the bounded bandwidth time duration product (Section 3.3), whereby a function cannot be both time and frequency limited. Any realizable impulse response, necessarily of finite duration, will thus have to be a truncated version of the theoretical impulse response. Such a truncation will degrade the actual FRF, the truncation in the time domain causing oscillations and spreading in the frequency domain (see Section 3.1, the Gibbs phenomenon).

To minimize the above effect, a multiplying window is applied to the truncated impulse response. The method is depicted by Figure T5.7, and summarized as:

(a) An ideal FRF H_{id} is specified.
(b) Its ideal impulse response h_{id} is obtained by an inverse Fourier transform,

$$h_{id} = F^{-1}[H_{id}] \tag{5.5a}$$

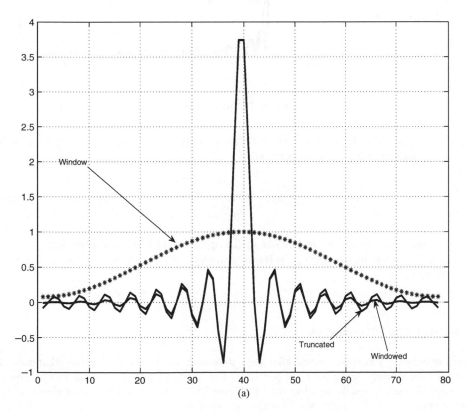

(a)

Figure T5.7

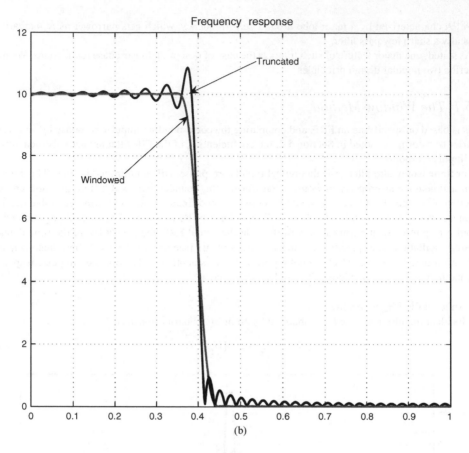

Figure T5.7 (*continued*)

For finite calculation h_{id} must be truncated. This is equivalent to multiplying it by a rectangular window, resulting in $h_{truncated}$ (Figure T5.7a):

$$h_{truncated}(n) = \begin{cases} h_{id}(n) & n \le n_{max} \\ 0 & n > n_{max} \end{cases} \qquad (5.5b)$$

(c) This discontinuity would generate oscillation (i.e ripples) in the FRF, the Gibbs effect.
(d) A suitable window, for example w_h, the Hanning window, multiplies h_{id} to avoid the discontinuities, resulting in

$$h_h = w_h * h_{id}$$

which equal the FIR parameters.

Various window functions, named after their developer, are in use – Hanning, Hamming and Kaiser windows being among the more prevalent. The filter's frequency response will be given by the convolution of H_{id} and W, the window's FT (Figure T5.7b). The ideal FRF can obviously only be approximated.

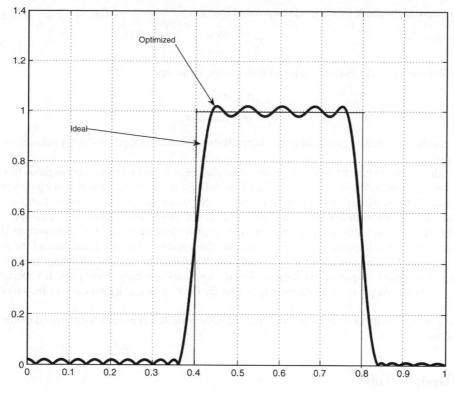

Figure T5.8

5.5.2 *Optimization-based Methods*

Other methods are based on optimizing a desired FRF, minimizing the error (see Figure T5.8) (and sometimes the MS error) between the desired and obtained function. In principle, more complex FRFs, of arbitrary shape, can be specified by these methods. Very popular are those using the so-called Remez algorithm in order to achieve the optimized solution.

5.6 The Importance of Linear Phase Filters

We first analyze a simple system consistent of pure delay τ. The impulse response is then

$$h(t) = \Delta (t - \tau)$$ (5.6a)

The corresponding FRF, computed as a Fourier transform, is

$$H(f) = \exp(-j2\pi f \tau)$$
$$|H(f)| = 1$$ (5.6b)
$$\phi_H = -2\pi f \tau$$

The negative of the phase with respect to the radial frequency

$$T_g = -\frac{d\phi(\omega)}{d\omega}$$
(5.7)

is called the group delay. For a phase proportional to the frequency

$$\phi = \alpha\,\omega$$

$$T_g = \alpha = \text{constant}$$

A system characterized by a constant group delay exhibits an actual real physical delay between output and input.

A nonlinear phase characteristic causes a so-called phase distortion in the signal propagation through it. Frequency components of the signal will each be delayed by an amount which is not proportional to frequency. Phase distortion is undesired in many applications, for example in music transmission. Another case is that of measuring multichannel transients. Linear phase filters will keep the delay between the transient's time location invariant. Linear phase FIR filters are thus an important class. Basically they utilize imposed symmetries in the impulse response. The details are beyond the scope of this chapter. Suffice to mention the four types of symmetries that exist (positive, negative, and the length of the impulse response, odd or even). Specific applications – high or low pass, differentiators, etc. – may dictate the use of one of the four types, with the limitations usually indicated by most modern software.

Using readily available filter design software, the effort involved is not much different than for any other filter.

5.7 Design Tools

Modern software, often with very convenient graphical interfaces, enable unspecialized users to design and implement such filters. It would be inpractical to describe any specific one, as more efficient, powerful and friendly ones appear all the time. What is important is that their use is so convenient *that there is little difference in the necessary effort to design simple or more elaborate filters* from the classes mentioned above.

The inputs needed would often be:

- General class – low pass, high pass, arbitrary FRF, linear phase, etc.
- Type – (IIR,FIR)
- Specifications – frequencies, attenuations, transition ranges.

Knowing the general properties of the main available filters and design approaches, it is usually possible to make a reasonable choice of the specific filter to be designed and used.

As stated in the objectives, the emphasis in this chapter is toward the user of such filters. Typically the processing would be done offline on acquired data. It should be stressed that *realization and implementation* of filters, for example programming a special purpose DSP, is a completely different problem, necessitating an in-depth familiarity with all filter aspects, not covered at all by this text.

6

Time Domain Averaging (Synchronous Averaging)

This is a method of extracting periodic signals from a composite signal, based on averaging signal sections of the period sought (see Braun, 1975, 1986). Knowledge of the frequency (or period) sought is necessary.

6.1 Principle

The principle is shown in Figure T6.1, whereby signal sections separated by one period are averaged.
Formally

$$y(n\Delta t) = \frac{1}{N} \sum_{r=0}^{N-1} x(n\Delta t - rM\Delta t) \tag{6.1}$$

with M the number of elements per period and N the number of sections averaged. $y(n)$ is then a sequence of M points, spanning one period of the averaged sections.

The frequency response can be computed via the Z transform as

$$Y(z) = \frac{1}{N} \sum_{r=0}^{N-1} (1 + z^{-rM}) X(z)$$

$$H(z) = \frac{Y(z)}{X(z)} = \frac{1}{N} \frac{1 - z^{-MN}}{1 - z^{-M}} \tag{6.2}$$

The frequency response is then

$$\left| H(f/f_p) \right| = H(z) \Big|_{z=\exp(-j\omega\Delta t)} = \frac{1}{N} \frac{\sin(\pi N f/f_p)}{\sin(\pi f/f_p)} \tag{6.3}$$

with

$$f_p = \frac{1}{M\Delta t}$$

the frequency of the extracted periodic component.

Discover Signal Processing: An Interactive Guide for Engineers   S. Braun
© 2008 John Wiley & Sons, Ltd

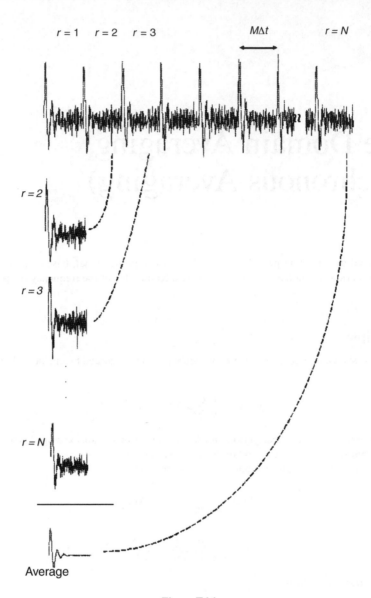

Figure T6.1

The FRF is shown in Figure T6.2. It has a form of a 'comb' filter, with main lobes centered around integer multiples of the synchronizing frequency f_p. Thus it is ideal for extracting the fundamental as well as all harmonics of the signal, hence the periodic signal itself.

Increasing the number of averages N will generate more sidelobes, all lobes will become narrower, with the bandwidth of the main lobe being inversely proportional to N. One of the possible definitions of the bandwidth around each main lobe is the equivalent noise bandwidth, the ratio of the area of the square of H and the maximum gain in the pass band (which in this case equals 1):

$$\mathrm{BW_{EB}} = \int_{-0.5}^{0.5} \left[\frac{1}{N} \frac{\sin(\pi N x)}{\sin(\pi x)} \right]^2 \mathrm{d}x = \frac{1}{N} \tag{6.4}$$

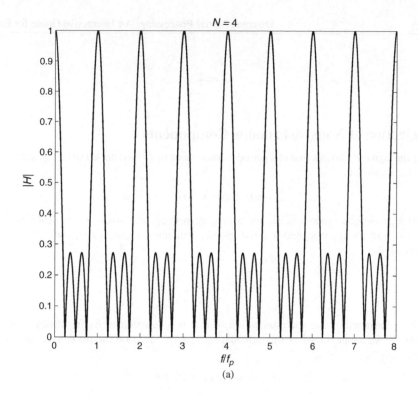

(a)

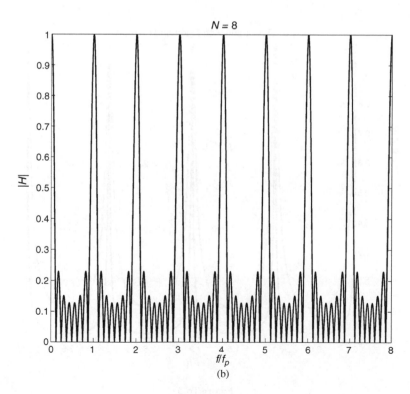

(b)

Figure T6.2

where

$$x = \frac{f}{f_p}$$

6.2 Rejection of Nonsynchronous Components

Denoting the signal $x(n\Delta t)$ as a sum of a periodic component $s(n\Delta t)$ and noise $e(n\Delta t)$, the signal elements not periodic within $1/f_p$

$$x(t) = s(t) + e(t) \tag{6.5}$$

For additive broadband random noise, the averaging process will attenuate the RMS of the noise, as the noise samples are independent.. For independent noise samples $e(t)$, their RMS will equal (asymptotically)

$$\bar{e}_{RMS} = \frac{1}{\sqrt{N}} e_{RMS} \tag{6.6}$$

For additive harmonic noise, the attenuation of $e(t)$ is a function of its frequency (according to Figure T6.2). An upper bound of this attenuation can be based on the maxima of the secondary lobes, resulting in

$$H_{max}(k) = [N \sin(\pi f/f_p)]^{-1} \tag{6.7}$$

where k denotes the lobe's index.

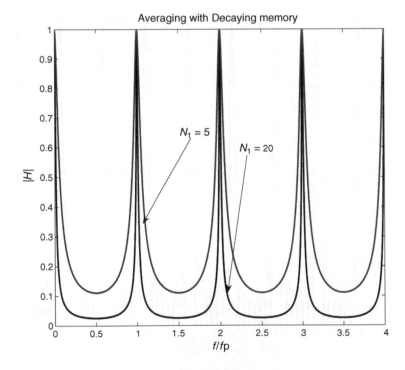

Figure T6.3

It can be noted that the increase in the attenuation of the noise with N has an oscillatory character: when N changes, the side lobes will be shifted compared to the interfering frequency. Increasing N will decrease the additional noise in an oscillatory manner, decreasing asymptotically.

Using Equations (6.3), we can summarize by finding the number of averages N, necessary to achieve a given attenuation, given by

Broadband noise: $N > (1/NR)^2$

Harmonic noise: $N > [NR \sin(\pi f/f_p)]^{-1}$ (6.8)

where NR is the desired noise reduction of the interfering signal components.

6.3 TDA with Decaying Memory Process

A recursive form of Equation (6.1) is given by

$$y_r(n\Delta t) = y_{r-1}(n\Delta t) + \frac{x_r(n\Delta t) - y_{r-1}(n\Delta t)}{r} \qquad (6.9)$$

where y_r is the running average at the rth period. If we use a fixed parameter N_1 instead of the running r

$$y_r(n\Delta t) = y_{r-1}(n\Delta t) + \frac{x_r(n\Delta t) - y_{r-1}(n\Delta t)}{N_1} \qquad (6.10)$$

the frequency response again has the form of a comb filter, but only one lobe exists around each center frequency. This is depicted by Figure T6.3. The proof is given in Braun (1986). Their bandwidth dictated by the parameter N_1.

The bandwidth is fixed, hence no additional interference rejection occurs after the transient response of the filter has stabilized. This is typical of systems with a decaying memory, not remembering the distant past. Often a low pass RC filter, composed of a capacitor and resistor, is used to introduce such a concept. For such a case, the memory decay is approximately exponential.

7

Spectral Analysis

7.1 Introduction

Spectral analysis is a signal processing method aimed at representing the dynamic patterns of signals in the frequency domain. To some extent it is a different, complementary way of looking at data (Bendat and Piersol, 1993; Braun, 1986).

This chapter deals with spectral analysis based on Fourier methods, all computations being based on the DFT. The objective is to familiarize the reader with the following aspects:

- Choice of the analysis tool and EU, according to the signal class
- Analysis errors and how to control them
- Implication of the above theoretical aspects for practice.

7.1.1 Overview

The basic analysis tool is the DFT, to be computed via the FFT. Specific spectral representations can be geared to specific signal classes:

- Fourier series would describe periodic data.
- Fourier transforms would describe transients.
- Power spectral density would describe random power signals.

All representations are derived via the basic DFT, and computed via the FFT. For quantitative results, the different EU used for these analyses are important.

Accuracy of the spectral analysis is addressed according to different error mechanisms:

- Leakage of computed spectral power into regions not spanned by physical signal
- Bias errors, systematic errors usually underestimating spectral peaks
- Random errors.

These errors can be controlled by appropriate tools. Leakage is controlled by windowing the data. Bias errors are controlled by choosing a long enough signal duration, so as to compute with sufficient frequency resolution. Random errors are controlled by averaging spectra of adjacent signal sections. Asymptotic expressions enable us to estimate the achievable random error.

Discover Signal Processing: An Interactive Guide for Engineers   S. Braun
© 2008 John Wiley & Sons, Ltd

A practical approach is often needed, considering the actual physical signals and limitations. Composite signals, not necessarily of a single class, may exist. Trade-off between bias and random errors exists for finite duration data.

7.2 Representation of Signals in the Frequency Domain

The basic tool will be the DFT, as defined for two discrete sequences $\{x(n)\}$, n = 0, $N - 1$ and $\{X(k)\}$, k = 0,$N - 1$:

$$X(k) = \sum_{n=0}^{N-1} x(n) \exp\left(-j\frac{2\pi}{N}ik\right)$$

$$X(n) = \frac{1}{N}\sum_{k=0}^{N-1} X(k) \exp\left(j\frac{2\pi}{N}ik\right)$$

(7.1)

The actual tool used will often be signal dependent. We thus address the spectral representation as adapted to different types of signal classes, this being of utmost importance whenever EU (engineering units) are to be computed. Engineering applications usually utilize one-sided spectral representations. The basic FFT, however, computes two-sided transforms, hence the emphasis on both presentations in the following sections.

It is of interest to state again the uncertainty principle, whereby the resolving power in the frequency domain, i.e. the possibility of separating components according to frequency, is limited by the signal duration (Section 3.3). The frequency spacing in the DFT is

$$\Delta f = \frac{1}{N\Delta t}$$

which is the reciprocal of the signal duration $T_{total} = N.\Delta t$, and is in accordance with the uncertainty principle. It would make no sense to strive for a smaller frequency spacing, in view of the impossibility of separating closer components.

7.2.1 Periodic Signals

Basically these are represented by Fourier series. One- and two-sided representations can be used, as described in Section 3. The Fourier series components are computed as

$$X_{FS}(k) = \frac{1}{N}X(k) \quad \text{Two sided}$$

$$X_{FSonesided}(k) = \begin{cases} \dfrac{2}{N}X(k) & k = 1... \dfrac{N}{2} - 1 \\ \dfrac{1}{N}X(k) & k = 0, \ K = \dfrac{N}{2} \end{cases}$$

(7.2)

The units are obviously those of $x(n)$, and the frequency scale is

$$f(k) = k\Delta f$$

Figure T7.1 shows an example of a periodic, decaying oscillating signal, Figure T7.1b showing one period. We note a fundamental component, and harmonics with frequencies being an integer multiple of the fundamental frequency (Figures T7.1c and d). The spectrum of periodic signals is often called a line spectrum.

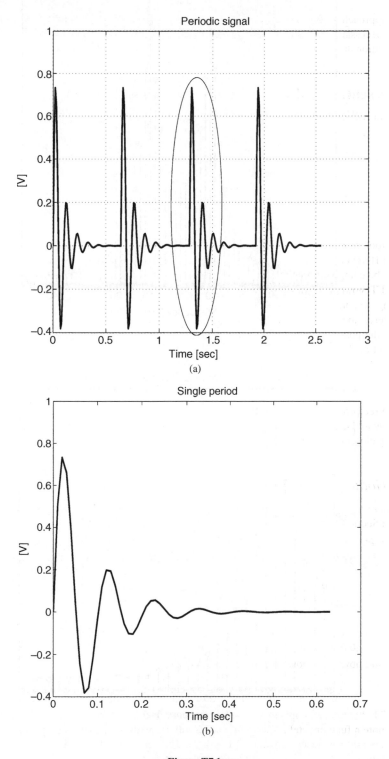

Figure T7.1

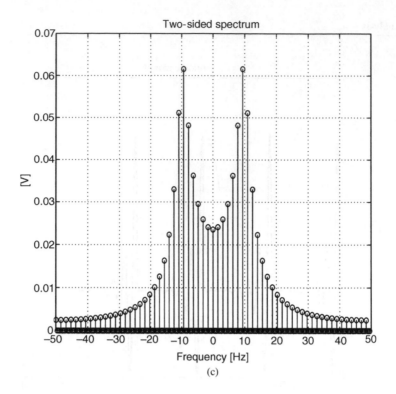

(c)

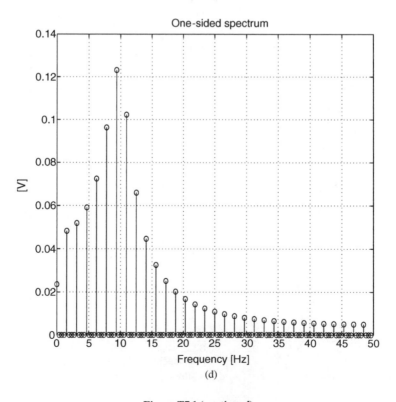

(d)

Figure T7.1 (*continued*)

We have chosen a case where the signal analyzed spans p integer number of periods, and the more general case will be deferred to Section 7.3.1, addressing the phenomenon of leakage. For this case (see Figure T7.2), we note that the location of the fundamental occurs at an index of $k = p$, i.e.

$$f_0 = k\Delta f$$

7.2.2 Transient (Aperiodic) Signals

An example is shown in Figure T7.3. A Fourier transform is to be applied. The spectrum is obviously continuous – but *computed* results can only be given for discrete frequencies. As addressed in Chapter 3, the straightforward application of the FFT assumes a normalized sampling interval $\Delta t = 1$. For results with engineering units we use Equation (7.3). The units are x-sec.

$$X_{\text{FTtwosided}} \frac{\Delta t}{N} X(k) \quad \text{Two sided}$$

$$X_{\text{FTonesided}}(k) = \begin{cases} \dfrac{2\Delta t}{N} X(k) & k = 1... \dfrac{N}{2} - 1 \\[2mm] \dfrac{\Delta t}{N} X(k) & k = 0, \ k = \dfrac{N}{2} \end{cases} \tag{7.3}$$

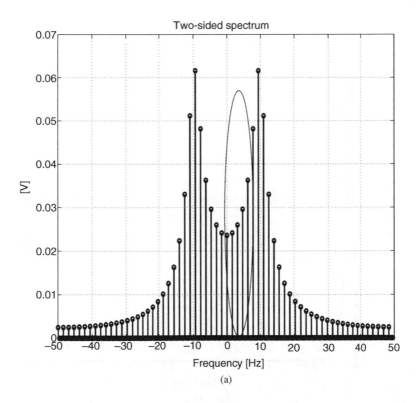

(a)

Figure T7.2

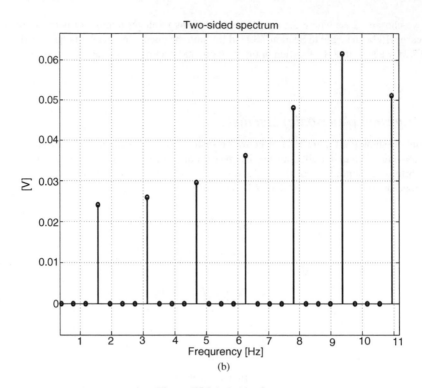

(b)

Figure T7.2 (*continued*)

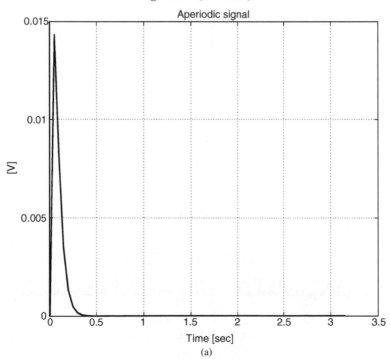

(a)

Figure T7.3

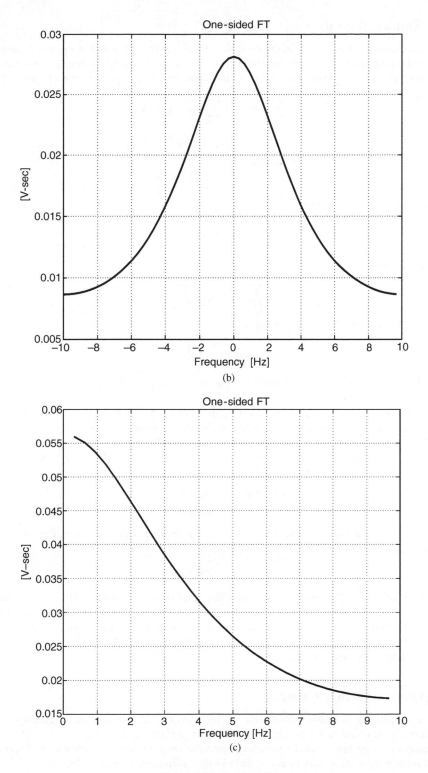

Figure T7.3 (*continued*)

7.2.3 Random Signals

The description of random signals in the frequency domain is usually done via a power density distribution (PSD) function. One possible definition of the PSD is via a general distribution function S, such that its integral (sum, for the discrete case) over the frequency range equals the signals power P. Thus

$$P = \sum_{k=0}^{N-1} S(k)\Delta f = \sum_{k=0}^{N/2} S_{\text{ONE SIDED}}(k)\Delta f$$

$$S(k)_{\text{ONE SIDED}} = \begin{cases} 2S(k) & k = 1... \quad \dfrac{N}{2} - 1 \\ S(k) & k = 0, \ k = \dfrac{N}{2} \end{cases} \tag{7.4}$$

The PSD itself is readily computed via the FFT.

$$\frac{1}{N\Delta t} \sum x^2 \Delta t$$

must equal the total power

$$\sum S \Delta f = \frac{1}{N\Delta t} \sum S$$

Substituting the Parseval law (Equation 3.11)

$$S(k) = \frac{\Delta t}{N} |X(k)|^2 \tag{7.5}$$

But the signal being random, this is applied to only one realization of the signal (see Exercise 2.3). The units of S are $x^2/\text{Hz} = x^2\text{-sec}$.

Correlation functions of signals were described in Section 2.4. These and the PSD functions cannot be unrelated, as they give information, in different domains, of signal patterns. A more basic definition of the PSD then exists via the Wiener–Khintchine theorem, whereby the autocorrelation and the PSD are related via a Fourier transform as per Equation (7.6) (Blackman, 1958). The computation as per Equation (7.5), utilizing the FFT, is now almost universal.

$$S(f) = F\left[R(\tau)\right] \qquad R(\tau) = F^{-1}\left[S(f)\right]$$
$$S(f) \longleftrightarrow R(\tau) \tag{7.6}$$

As an example, let us take a 'white noise', described as having $S(f) = $ constant over the complete frequency range. From Equation (7.6) we get the result

$$R(\tau) = P_t \delta(0)$$

with P_t the total power. The autocorrelation of white noise is a delta function. This can be considered as a definition of white noise, the correlation between any two separated times being zero, implying no internal order or pattern. A white noise is of course a mathematical entity only, as it implies infinite power (constant PSD over an infinite frequency range). In reality signals can have white noise properties only in a finite frequency range.

7.3 Errors and their Control

The ease of utilizing available software (or instruments) to compute spectra of signals may result in overlooking the important aspect of uncertainties in the results. It is necessary to understand the mechanism by which errors can be introduced into the computations, and even more important, to understand how these errors may be controlled to stay within predescribed limits.

Three major types of errors are encountered in the area of spectral analysis:

(a) Leakage errors
(b) Bias errors
(c) Random errors.

Leakage is a phenomenon by which spectral energy, located physically (i.e. from the actual true physical behavior) in a specific range, appears, due to the computations, to affect other frequency regions.

Bias errors are systematic errors, over- or underestimating the correct results. Random errors cause an uncertainty in the results, to be quantified by statistical parameters. The variance or standard deviation of this error can be used to define an uncertainty gap around the computed results.

A more fundamental error mechanism, common to all digital signal processing tasks, is the aliasing error. Such an error is introduced by incorrect sampling, and cannot be controlled when performing spectral analysis of already digitized data. It is hence assumed that the aliasing error has been avoided by correct sampling (see Chapter 10).

7.3.1 Leakage Errors

As discussed in Section 3.5, periodicities are introduced to discrete signals when a DFT operation is applied. The section of the signal analyzed is actually repeated periodically outside the analyzed time window. Discontinuities may thus occur at the start and end of our signal section.

The effect of such a discontinuity is easy to demonstrate for a harmonic signal, as in Exercise 3.12, and shown again in Figure T7.4. The only case when the discontinuity disappears is when an integer number of periods exists in the time section analyzed. The effect of the discontinuity appears as many

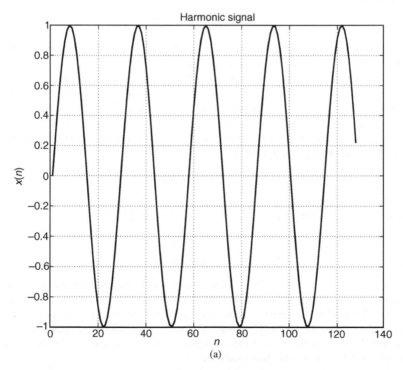

(a)

Figure T7.4

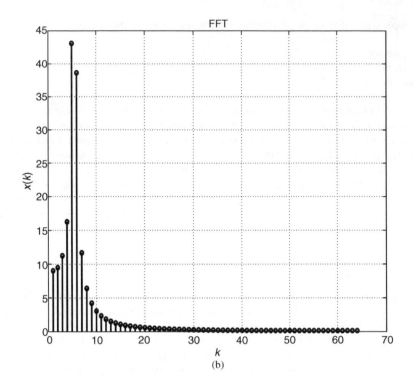

(b)

Figure T7.4 (*continued*)

additional frequency components in the spectrum. Leakage can thus occur whenever the analyzed signal is not periodic within the analyzed signal section.

The discontinuity can be reduced by applying a so-called window to the signal. The windowing operation consists of artificially giving less weight to the end points. The weight must obviously be gradual in order not to generate other types of discontinuities. The window function is centered around the middle of the signal section, gradually going to zero at the extremes. This is shown in Figure T7.5. The weighting is given by Equation (7.7), with $x(n)$, $w(n)$ and $x_w(n)$ the original signal, the window function and the windowed signal respectively.

$$x_w(n) = x(n)w(n) \qquad (7.7)$$

To understand the effect of windowing, we note that analyzing the original unwindowed signal is actually equivalent to using a rectangular window w_r, defined by Equation (7.8):

$$w_r(n) = \begin{cases} 1 & 0 \leq n \leq N-1 \\ 0 & \text{otherwise} \end{cases} \qquad (7.8)$$

The effect of applying such a rectangular window can be seen in the frequency domain, by applying a Fourier transform to Equation (7.7), resulting in the convolution of the transforms of $w(t)$ and $x(t)$, (Equation 7.9). The window has $\sin(x)/x$ character, hence a main lobe, and secondary lobes whose peak decreases slowly with frequency. This shape of $W_r(\omega)$ results in a spread spectrum, in spite of the line spectrum of the original harmonic function. Leakage thus occurs, as energy seems to appear at frequencies different from that of the line spectrum.

$$x(t) \longleftrightarrow X(\omega) \otimes W(\omega) \qquad (7.9)$$

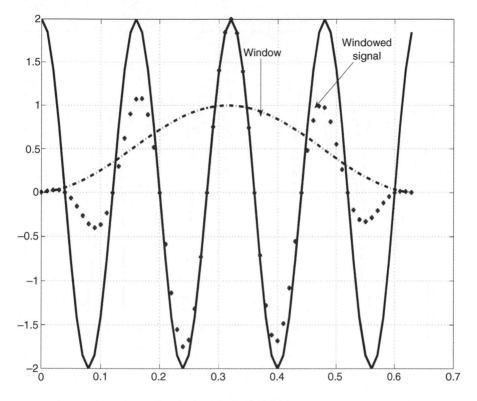

Figure T7.5

Improved window characteristics can decrease the spread, and hence the leakage. This is shown in Figure T7.6. A multitude of windows exist, each one with some specific characteristics. Most of them will have secondary lobes whose peaks decrease much faster with frequency, but at the price of some widening of the main lobe. There is usually a trade-off between these two characteristics.

One of the most popular windows used in order to control leakage is the Hanning window, w_h, given by Equation (7.10). The side lobes have a roll-off slope of 18 dB/octave as compared to 6 dB/octave for the rectangular one (see Figure T7.6). The central lobe of the Hanning window is wider, however.

$$w_h(n) = 0.5 - 0.5\cos\left(\frac{2\pi n}{N-1}\right) \tag{7.10}$$

The application of a window (except the rectangular one) reduces the total energy/power of the signal. A correction factor needs to be applied in order to minimize this effect. Most commercial software will apply such a correction automatically. Unfortunately the correction is never ideal, as the energy/power reduction will also depend not only on the window type chosen, but also on the specific signal analyzed.

7.3.2 Bias Errors

Bias errors are systematic errors. In spectral analysis, they are caused by a computationally insufficient frequency resolution.

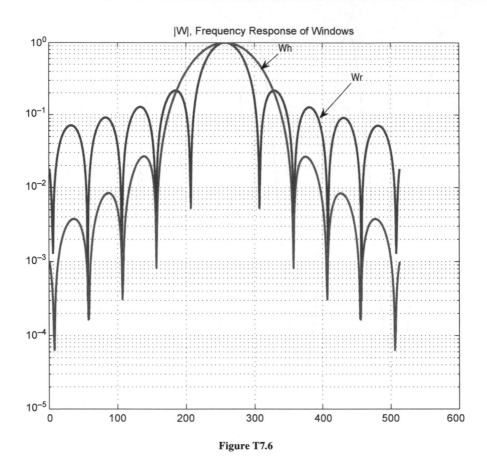

Figure T7.6

The frequency steps at which the spectra are computed are discrete, given by

$$\Delta f = \frac{1}{N \Delta t}$$

It is obviously impossible to see local changes whose separation is less than $2\Delta f$. This is analogous to looking at any function with a measuring scale whose resolution is too coarse to see details smaller than the minimum resolution of this scale. It is even possible to understand qualitatively the type of error introduced: local peaks will be underestimated, while local minima will be overestimated (see Figure T7.7).

Another specific example is shown by Figure T7.8. The actual signal is composed of the sum of two decaying oscillating signals of close frequency, and the interference can be seen from the figure. A computational resolution of approximately $\Delta f = 1/T_t$ results from analyzing a signal of 1.6 sec duration, and two close peaks can indeed be seen in the frequency domain. When the computational resolution, the frequency step, is now doubled by using half the signal length by shortening the signal length by factor of 2, we lose the capacity to separate the two peaks.

The bias error described above was tested for a deterministic signal, showing the underestimation of spectral peaks. Similar considerations also occur for random signals: spectral peaks will be underestimated (as in Figure T7.7) unless the computational frequency step is at least three to four times smaller than any width in the true spectrum.

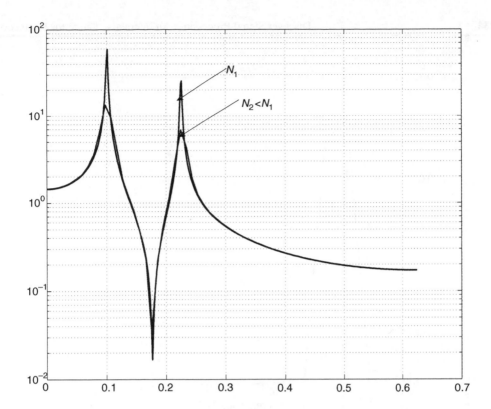

Figure T7.7

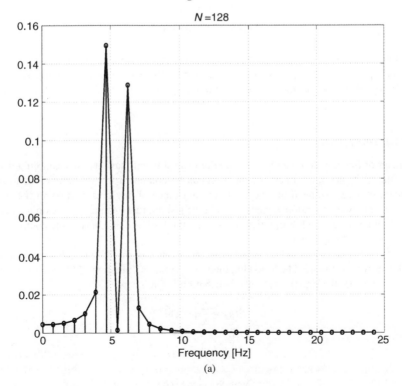

Figure T7.8

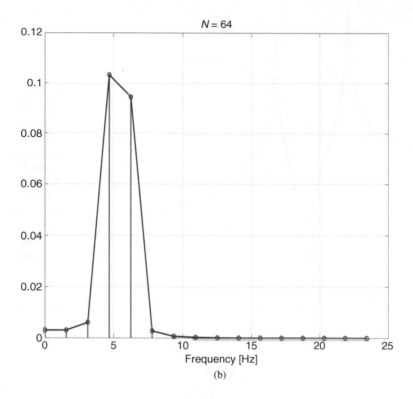

Figure T7.8 (*continued*)

An exact analytical expression for the bias error can only exist for specific analytic signal forms. A more practical approach is to require a Δf step at least three times smaller than any separation required. The problem of course is that the required separation is not known! A practical approach will be described later.

7.3.3 Random Errors

Any parameter or function computed from a random variable will have its own sampling distribution. Only an estimate of such a parameter or function can be computed from a finite realization. From a practical point of view, we need to investigate the existence of any bias or random (i.e. variability) errors. Spectral analysis of random signals is of course just another example where only an estimate of the true PSD can be calculated. Properties concerning the variance and bias of the computed estimate are of utmost practical importance.

For a single realization of a random signal $x(n)$, an estimate of the PSD is given by Equation (7.11), where the hat denotes estimate. The following theoretical results, Equation (7.12), are then obtained for the specific case where the signal amplitude distribution is Gaussian.

$$\hat{S}(k) = \frac{\Delta t}{N}|X(k)|^2 \tag{7.11}$$

$$E\left[\hat{S}(k)\right] = S(k) \tag{7.12a}$$

$$\text{Var}\left[\hat{S}(k)\right] = S^2(k) \tag{7.12b}$$

$$\sigma\left[\hat{S}(k)\right] = S(k) \tag{7.12c}$$

$$e_r[\%] = 100 \,\frac{\hat{S}(k)}{S} \tag{7.12d}$$

Equation (7.12a) shows that the estimate is unbiased. Equation (7.12b), however, shows an unacceptable variability: the relative random error (RMS error, that is) is 100%! The result is due to the fact that each point of the PSD function is computed as the sum of two square values, the squares of the real and imaginary part of the DFT. It has a chi square distribution with 2 degrees of freedom, for which the results of Equation (7.12) apply. It can be noted that the number of points does not appear in the equation. In Section 2.4 the concept of random errors in estimating parameters and functions was introduced. Increasing the number of data points available would often reduce the random error. As far as spectral analysis of random signals is concerned, the variability does not decrease with N ! One intuitive explanation is that the number of spectral values to be estimated also increases with N, thus the number of spectral results per data points is unchanged.

An example is shown by Figure T7.9, showing the superposition of the PSD of 10 realizations for various values of N. The enormous variability is obvious. Also shown in the figure is the average (for

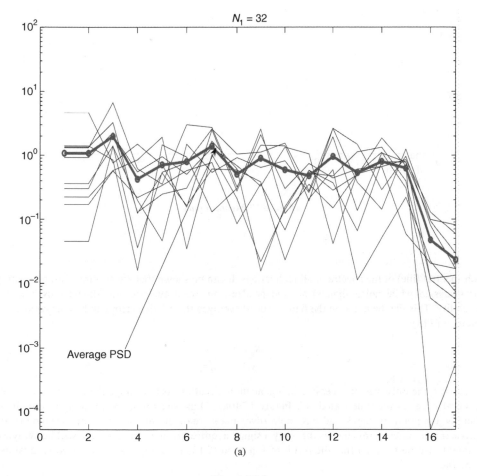

Figure T7.9

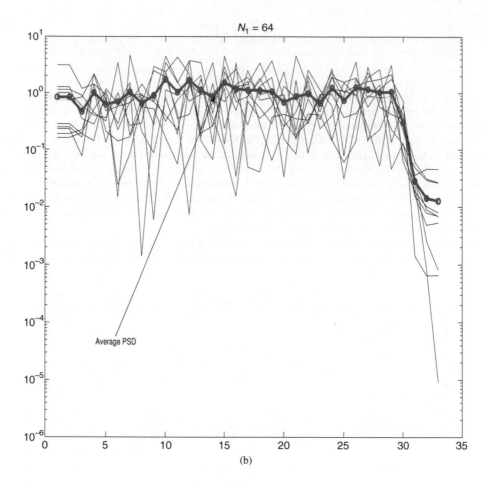

(b)

Figure T7.9 (*continued*)

each spectral line) of the spectra of all realizations. It can be shown that the spectra of all realizations (for Gaussian white noise signals) are independent, an averaged spectrum should thus have less variability. This can be noted in the figure. For M averages the relative error can be computed as per Equation (7.13):

$$e_r = \frac{S_M(k)}{S(k)} = \frac{1}{\sqrt{M}} \tag{7.13}$$

In practice, the only way of accessing independent realizations is to average the spectra of different temporal slices of the time signal (see Figure T7.10a). Thus we assume the property of ergodicity, whereby temporal averages and ensemble (over many realizations) averages are asymptotically equivalent. In practice this means that only a signal acquired from a single source will be analyzed. It should also be stressed that the result of Equation (7.13) is asymptotic, to be used at most as a guideline.

The computational principle, showing how the signal is segmented for the computations, is shown in Figure T7.10(a). The number of segments used for averaging can be increased by overlapping segments, as per Figure T7.10(b). The segments need to be windowed, say by a Hanning window, in order to give less weight to the same data points appearing in adjacent segments (see Figure T7.10c). Practical overlapping is limited to 25–50%, resulting in an additionally reduced random error. Overlapping is mostly effective for wideband random signals.

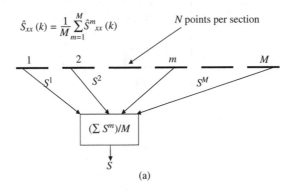

(a)

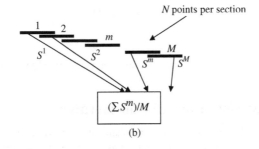

(b)

Overlapping

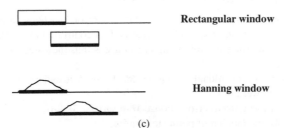

(c)

Figure T7.10

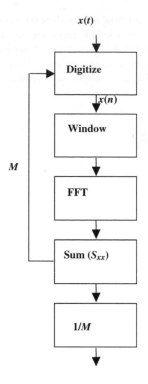

$x(t)$

Digitize

$x(n)$

Window

M

FFT

Sum (S_{xx})

$1/M$

Figure T7.11

The algorithm, enabling us to control the random error, is described by the flow chart of Figure T7.11.

7.4 Spectral Analysis: Practical Considerations

Having discussed basic algorithms and error mechanisms, we briefly address practical aspects. The parameters of the analysis have to be specified, so as to obtain the desired accuracy. This will often necessitate both judgments and compromises.

7.4.1 Controlling Errors for Random Signals

Using the method of segmenting, we can control the bias error by choosing a large enough N, the number of points in each segment. The random error is then controlled by M, the number of sections whose spectra are averaged. The basic following procedure is thus indicated:

(a) Specify the necessary computational resolution, Δf, choose N to satisfy $N = 1/\Delta f \Delta t$.
(b) Approximate N to a power of 2.
(c) Specify the random (normalized) error, choose M to satisfy $M = 1/\sqrt{(e_r)}$.
(d) Choose $N_{total} = N*M$, the number of points to acquire.

This procedure shows the importance of specifying performance before acquiring any data! A more problematic case may occur if the number of available data points, N_{total}, is fixed, for example if the signal is already acquired. Then the bias and the random error may need be to be balanced. Decreasing the bias error by increasing N results in a reduced M, an increase in the random error, and vice versa. Some practitioners feel that bias errors are more important, as underestimating peaks could be extremely undesirable. Thus N could be increased until no increase in peaks could be recognized. This may be difficult to determine exactly, due to the increasing random error, but a reasonable choice is often possible.

Some possible improvement can be achieved by so-called overlapping processing, a topic beyond the scope of this text.

7.4.2 Leakage and Windows

The fact that spectral energy has leaked to other frequency ranges may obscure the existence of weak components in the said region. This leakage can completely obscure the existence of some components, unless we attempt to control this effect by applying a window. Leakage can decrease the dynamic range, the ratio of the maximum to the minimum amplitude of components which can be recognized. Applying a window, and thus decreasing the leakage, can thus increase the dynamic range of the spectral analysis. Exercise 7.4 deals with such a specific case.

The application of windows seems indicated whenever a discrete spectrum with closely spaced lines and huge dynamic range is predicted. Often the window is applied as a default. The so-called Welch method, based on segmenting and averaging the data, usually applies a window by default.

7.4.3 The Analysis of Combined Signals

The classification of signals into deterministic periodic or transient ones, or random ones, is important in order to clarify the major aspects in their analysis. In practice, however, we often find that a signal is composite, including more than one type of signal class. Any experimental data acquisition includes some random noise, and sometimes harmonic interferences of frequencies which are integer multiples of the line frequency.

Some qualitative guidelines are possible. The existence of random components necessitates some segment averaging. It is thus a matter of some judgment to decide on the number of averages needed. For signals which are basically deterministic and periodic, with a large signal to noise ratio, a small number of averages may suffice. For very small signal to noise, the number of sections M could be more or less as if the signal was of purely random character. No averaging is of course attempted for transient deterministic signals, unless it could be acquired repetitively.

The question also arises of what type of frequency presentation should be computed. In section 7.2, we discussed Fourier series, integrals and PSD. We often use the PSD as a general representation. For periodic signals, the power being concentrated in discrete lines, the PSD must include impulse functions.

In practice, as all computation use the DFT as computed by the FFT. The difference in the computational approach for different signal classes may not be of utmost importance, thus the popularity of using the PSD. When only a qualitative result is needed, where the spectrum shape is of major interest, the engineering units are unimportant, and only consistency in the analysis needs to be maintained. It is of course the interpretation which is the most important aspect.

7.4.4 Frequency Resolution and Zero Padding

The duration of signals (and especially transients) is usually dictated by the physical situation. The number of data points to be acquired may, however, cover a section of larger duration, and hence include zero value data appended to the signal.

To see the effect of such zeros, we note the DFT of such a case (Equation 7.14), where $N - N_1$ zeros have been appended:

$$X(k) = \sum_{n=0}^{N-1} x(n)\exp\left(-j\frac{2\pi}{N}nk\right) = \sum_{n=0}^{N_1-1} x(n)\exp\left(-j\frac{2\pi}{N}nk\right) + \sum_{N_1}^{N-1} 0\exp\left(-j\frac{2\pi}{N}nk\right) \qquad (7.14)$$

Nothing is contributed to the DFT from the second term, thus the DFT value is unmodified by appending of zeros. The computational frequency step is however modified, as

$$\frac{1}{N_1\Delta t} > \frac{1}{N\Delta t} \quad \text{with} \quad N_1 < N$$

The resolving power of the DFT is limited by the uncertainty theorem. The apparent improved computational resolution is actually a special case of interpolating the DFT at additional frequency points. Exercise 7.5 deals specifically with this topic.

8

Envelopes

8.1 Introduction

Band-limited signals are an important class of signals in various domains such as communications, mechanical engineering and others. They are defined as signals whose spectrum is zero outside a frequency region of $\pm B$ around a central frequency f_0. We define narrow band signals as characterized by $B \ll f_0$. An example of such a signal is shown in Figure T8.1. They include signals which are modulated in amplitude or frequency. In the field of communications we encounter the AM (amplitude modulation) and FM (frequency modulation) radio waves. The information signal modulates a carrier signal in the transmission phase, whereas a demodulation is performed at the reception in order to extract back the information signal. In many other domains, measured signals are also physically modulated, and often the demodulation may allow better interpretation of the signals by extracting only the useful information in the signal. Typical examples are acoustic waves in structures, vibration in rotating machines, electric current in induction machines, and evoked potential in the biomedical field. Nowadays, this operation is done numerically (in discrete time operation), based on concepts like the Hilbert transform and of analytic signals.

8.2 The Hilbert Transform (HT) (Feldman, 2002)

The HT of a signal is a time domain signal defined by

$$H[x(t)] = \hat{x}(t) = \frac{1}{\pi} \int_{-\infty}^{\infty} \frac{x(\tau)}{t - \tau} d\tau = x(t) \otimes (1 / \pi t) \tag{8.1}$$

Applying a Fourier transform

$$F[\tilde{x}] = F[x(t)]F[1 / \pi t] = X(f)\text{sign}(f) \tag{8.2}$$

where sign(f) is the FT of ($1/\pi t$):

$$\text{sign}(f) = -1 \text{ for } f < 0; \qquad 0 \text{ for } f = 0; \qquad 1 \text{ for } f > 0$$

The HT has the effect of shifting negative frequency components of $x(t)$ by $+90°$ and those of the positive ones by $-90°$, and can be thought of as a $90°$ quadrature filter.

It is trivial to show that an inverse Hilbert transform can be obtained as

$$x(t) = -H[\hat{x}(t)]t \tag{8.3}$$

Discover Signal Processing: An Interactive Guide for Engineers   S. Braun
© 2008 John Wiley & Sons, Ltd

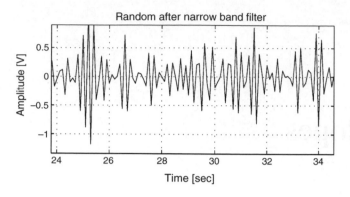

Figure T8.1

Applying an HT transform to the harmonic signals $\sin(2\pi f_0 t)$ and $\cos(2\pi f_0 t)$ results in

$$H[\sin(2\pi f_0 t)] = \cos(2\pi f_0 t)$$

$$H[\cos(2\pi f_0 t)] = -\sin(2\pi f_0 t)$$

A modern computational route is to compute the HT by calculating first the FT of $x(t)$, apply Equation (8.2). This then enables us to compute $\hat{x}(t)$ via an inverse Hilbert transform.

8.3 Analytic Signals

The analytic signal x_a that corresponds to the real signal $x(t)$ is a complex signal defined by

$$x_a(t) = x(t) + j\hat{x}t) \tag{8.4}$$

Thus $x_a(t)$ is complex with a real and imaginary part linked by the Hilbert transform. A convenient representation of the analytic signal is in the polar form

$$x_a(t) = A(t)\exp[j\phi(t)]$$

$$A = [x^2(t) + \tilde{x}^2(t)]^{0.5} \tag{8.5}$$

$$\phi(t) = \tan^{-1}\left[\frac{\tilde{x}(t)}{x(t)}\right]$$

and $A(t)$ is termed the *envelope* signal and $\phi(t)$ the *instantaneous phase* signal. An *instantaneous frequency* can then be defined as

$$f_{in} = \frac{1}{2\pi}\frac{d\phi(t)}{dt} \tag{8.6}$$

8.4 Narrow Band (NB) Signals and their Envelope

Such signals are centered around a frequency f_0, with $X(f)$ being zero for $f > f_{max}$, where $f_{max} \ll f_0$.

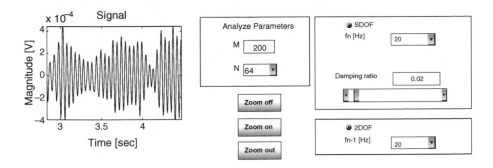

Figure T8.2

Using the analytic form of such a signal, the FT of it would be

$$X_{an}(f) = X(f) + j\tilde{X}(f) = \begin{vmatrix} 2X(f) & f > 0 \\ 0 & f < 0 \end{vmatrix} \tag{8.7}$$

and the analytic signal has spectral components only for positive frequencies. Multiplying the analytic signal by $\exp(-j2\pi f_0 t)$,

$$x_{lp} = x_{an} \exp(-j2\pi f_0 t)$$

and applying the shifting theorem of the FT (Equation 3.5), we interpret this as the x_{lp} being the complex envelope of x_{an} which is shifted from f_0 to zero frequency, and x_{lp} the envelope.

An NB signal can thus conveniently represented via the polar form of an analytic signal

$$x_a(t) = A(t) \exp[j\phi(t)] \tag{8.8}$$

with $A(t)$ the envelope, having low pass properties, i.e. being much slower than f_0.

Examples of random NB signals were encountered in previous chapters. For example, from Exercise 2.1, the (zoomed) plot shows such a signal (Figure T8.1).

Another example could be seen from exercises where a random excitation is applied to a second-order dynamic system with low damping, for example Exercise 7.6. This is depicted, after zooming, in Figure T8.2. In these examples, the envelope has a random (slow) character; the zero crossing rate is, however, almost constant, approximately the reciprocal of the center frequency of the pass band.

The instantaneous phase and frequency describe the fast fluctuations of the signal. The concept of an instantaneous frequency enables us to tackle situations where the frequency of a signal is time dependent. One example is the case of vibrations of rotating machines, where the rotational speed is not constant, for example start-up or coasting down. The speed variation can then be tracked by computing the instantaneous frequency of the vibration component due to the rotation.

9

The Spectrogram

9.1 Introduction

The time and frequency domain presentations are to some extent complementary, two different ways of looking at the same information. Using Fourier methods however, they are practically disjoint. The temporal occurrence of a specific event is lost in the FT (actually it resides in the phase information), and certainly in the PSD. Conversely, the frequency information cannot be localized to a certain time. Hence the need for time-frequency methods enabling us to analyze and interpret the time-varying spectral contents. Such methods find applications where the frequency of physical phenomena is not constant (rotating machines with nonconstant rotational speed, propagating waves in dispersive media where velocity is frequency dependent, speech/music, sounds emitted by vehicles/bats/whales, transient phenomena, etc.). Specific tasks could include:

(a) To estimate the instantaneous frequency of time-varying signal.
(b) When dealing with frequency-modulated signal, it is of interest to estimate precisely the frequency time variation law, for example, in a dispersive media.
(c) To localize precisely a short event in time and frequency. For example, the precise localization of an oscillating impulse may be of interest in many contexts.

The time frequency representations should be able to show a spectral representation as a function of time.

9.2 Time Frequency Methods

While this chapter deals with spectrograms (Qian, 1996; White, 2002), many additional methods are in use today. One classic possibility is to use a bank of filters with adjacent band pass regions, and to track the time changes of the filter outputs. Sometimes (especially for oscillatory signals of varying frequency), the instantaneous frequency and envelope functions are tracked via Hilbert transform based methods (Chapter 8). Others are beyond the scope of this text, and the following are merely listed:

- The Wigner Ville transform, a quadratic time frequency distribution, defined by

$$WV_x(t,f) = \int x\left(t+\frac{\tau}{2}\right) x\left(t-\frac{\tau}{2}\right) e^{-2\pi \cdot jf\tau} d\tau$$

- Wavelets – a decomposition of the signal $x(t)$ onto a family of basic functions in the timescale plane

Discover Signal Processing: An Interactive Guide for Engineers   S. Braun
© 2008 John Wiley & Sons, Ltd

or wavelets $h_{t,a}(u)$. t is the translation variable, a the scale variable.

$$WT_x(t,f) = \int x(u)h_{t,f}(u)du$$

$$h_{t,f}(u) = \left|\frac{f}{f_0}\right|^{1/2} h\left[(u-t)\frac{f}{f_0}\right]$$

In the classic Fourier methods, the decomposition is into an orthogonal basis of harmonic functions. In the wavelet method, the decomposition results in time-varying components, each one computed for different scales (inversely related to frequency).

9.3 The Short Time Fourier Transform (STFT) and the Spectrogram

The idea is to consider a nonstationary signal as a set of adjacent quasistationary signals. We consider a time-varying 'periodogram' obtained by sliding a window across the time signal (Figure T9.1). We then perform a Fourier transform. This may be written:

$$S_x(t,f) = \int_{-\infty}^{\infty} x(u)h(u-t)e^{-2j\pi ft}du \qquad (9.1)$$

where $x(t)$ is the signal to be analyzed, and $h(u-t)$ the sliding window. The expression given by Equation (9.1) is called the short time Fourier transform (STFT). The STFT has a fixed resolution in time and frequency, in the same manner as the Fourier transform. The frequency resolution $\Delta f = 1/N\Delta t$ is controlled by the length of the window $h(t)$.

The STFT modulus square

$$|S_x(t,f)|^2 \qquad (9.2)$$

is called a spectrogram. The spectrogram may be displayed in a 3D representation called a waterfall representation (Figure T9.2a) or in a 2D map representation (Figure T9.2b). The spectrogram is based

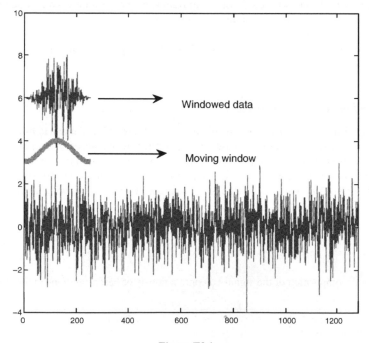

Figure T9.1

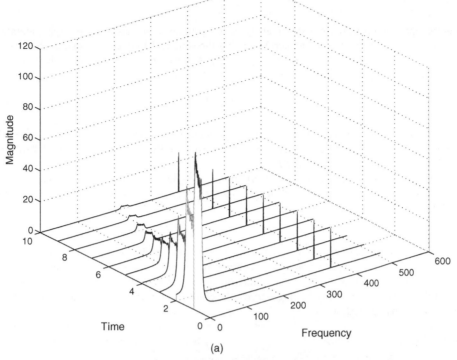

(a)

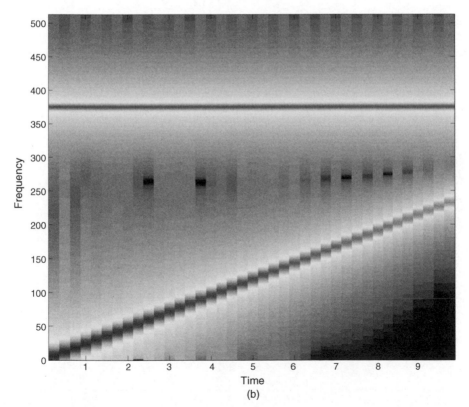

(b)

Figure T9.2

on a direct application of Fourier spectral analysis. Its limitations are thus of similar type, mainly due to the uncertainty principle (see Chapter 3) relating the time and frequency domain. This principle states that it is not possible to locate exactly a point (t, f) in the (T, F) plane. If we consider a signal $x(t)$ of duration T and frequency bandwidth B, then we have the inequality

$$BT \geq \frac{1}{2} \tag{9.3}$$

Thus to increase the resolution in the frequency domain (small B) we need a longer time signal (large T). Inversely, if we want to localize with precision in the time domain (small T) then the frequency resolution will suffer (large B).

Another aspect is that of linearity: the spectrogram, when analyzing a composite signal of say, two signals, is not the sum of the two individual spectrograms. As a simple example, let us take $y(t) = x_1(t) + x_2(t)$, then

$$S_y(t, f) = S_{x_1}(t, f) + S_{x_2}(t, f) + \text{Res} \tag{9.4}$$

The third term is the real part of the cross-spectrogram, and may be considered as an interference. This interference can be neglected only if the components of the x_1 and x_2 are sufficiently separated in frequency. This is in contrast to classic one-dimensional PSD presentations. It is more severe than with regular PSDs, where, for reasonably long records, the cross-spectra tend to zero, due to the orthogonal property of the decomposition.

10

Data Acquisition

10.1 Data Acquisition and Signal Processing Systems

Figure T10.2 shows a block diagram of a typical commercial measurement system which includes data acquisition and processing capabilities. The sensors (or sensors in a multichannel measurement application) are very often analog, i.e. generating a continuous voltage (usually) signal. This is passed through a signal conditioning unit, amplifying the signal to the input range of a digitizer, the digital to analog converter (ADC). Some prefiltering can be designed into the signal conditioner. A special purpose analog filter, known as an antialiasing filter (see later, Section 10.4) precedes the ADC. The sampled signal is then processed by the digital signal processor, which would have dedicated operations like spectral analysis, time domain (synchronous) averaging, etc.

The logical organization of the block diagram does not necessarily correspond to hardware subsystems. The sensor/signal conditioner unit could be separate, some signal processors would include the data acquisition module, or it could be a separate computing unit (even a general purpose one), etc. The point stressed here is that the data acquisition and signal processing phases must often be considered together, when designing or evaluating a signal processing task.

For digital signal processing, the signal must be discretized (quantized) in two domains: the time and amplitude domain.

10.2 Amplitude Quantization (Baher, 1990; Dally *et al.*, 1993)

The basic operation of a classic DAC is depicted by Figure T10.1. The amplitude domain is divided into equal bands. At the instant of sampling, the band within which the signal lies at this sampling instant is determined. The signal value can then be taken as one of the band limits.

The number of bands dictates the resolution of the quantization. This is a function of the number of bits of the ADC, usually a power of two. A 12-bit ADC would thus divide the amplitude range into $2^{12} = 4096$ possible values.

The measurement resolution will depends on two parameters: the range of the ADC, and N, the number of bits. For the above number of bits, an ADC with a full scale (FS) range of 0–10 [V].

$$\Delta V = \frac{V_{FS}}{2^N} \tag{10.1a}$$

$$\Delta V = 1/4096 = 244 \, \mu V$$

Discover Signal Processing: An Interactive Guide for Engineers   S. Braun
© 2008 John Wiley & Sons, Ltd

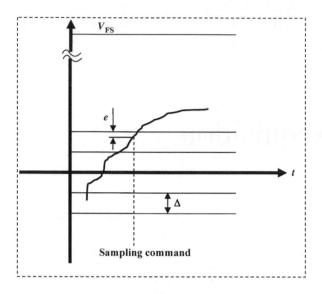

Figure T10.1

For bipolar input ranges, $N-1$ bits are available, as 1 bit is used for the sign, to have positive and negative ranges with half the voltage bands. For our case we would have 512 to +511 possible steps of

$$\text{(obviously the same)} \qquad 0.5/4096 = 1/2048 = 244 \ [\mu V]$$

The quantization introduces an uncertainty, known as the quantization error. The true signal could fall anywhere in the band at the moment of sampling. The difference is known as the quantization error, as per Figure T10.1. The upper bound is obviously $\pm 0.5 \ \Delta$ [V].

A more realistic error figure is to consider some mean error, assuming an equal probability of the true signal falling anywhere within the band. It can be shown that for such an assumption, the RMS value of the quantization error is

$$E_{RMS} = 0.29\Delta \ [V] \tag{10.1b}$$

The quantization error adds a noise to the signal. This is minimized by adapting the signal level applied to the ADC (by means of appropriate amplification), as shown by Exercises 10.1 and 10.2.

With modern data acquisition cards, the quantization error is not usually a problem, as 12-bit ADC are almost standard. Still it is important to recognize the effect. Matching the dynamic range of the signal to that of the ADC (through an appropriate amplification) can make this effect negligible

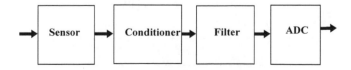

Figure T10.2

10.3 Quantization in Time: The Sampling Theorem (Ambardar, 2007)

Let us denote the continuous signal by $x(t)$, with an FT of $X(f)$. We assume ideal instantaneous sampling, yielding $x(n\Delta t)$. The FT of the sampled signal will exhibit a periodicity in the frequency domain as per Equation (10.2) (see Figure T10.3):

$$x_s = x(n\Delta t) = x(t)\left[\sum_{n=-\infty}^{n=\infty} \delta(t - n\Delta t)\right] \tag{10.2a}$$

Expanding the periodic term in the brackets to a Fourier series,

$$\frac{1}{\Delta t}\sum_{k=-\infty}^{k=\infty} \exp(j2\pi kt/\Delta t) \tag{10.2b}$$

Applying a Fourier transform to Equation (10.2b) and using the shifting theorem (Equation 3.5) results in

$$X_s = \frac{1}{\Delta t}\sum_{k=-\infty}^{k=\infty} X\left(f - \frac{k}{\Delta t}\right) \tag{10.2c}$$

where X_s is the FT of the sampled signal. X_s is shifted and repeated periodically at multiples of the sampling frequency.

Is it possible to have ideal sampling, where no information is lost? Such an ideal sampling, where all the information is retained in the discrete samples, would imply the possibility (at least theoretically) of computing, using some interpolation, the signal values between sampling times. The fact that overlapping of the periodic spectra occurs in some regions (Figure T10.3) would indicate that this is not generally possible.

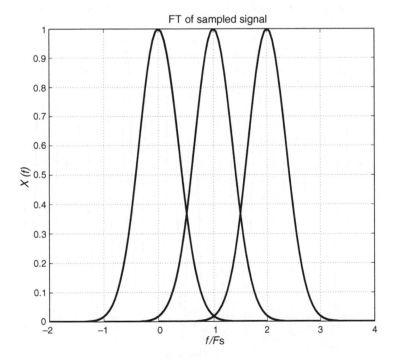

Figure T10.3

The sampling (also called the Nyquist) theorem indicates such a possibility for a specific case: that of a band-limited signal, when the spectrum of the signals is such that

$$S = \begin{cases} S(f) & f \leq f_{max} \\ 0 & f > f_{max} \end{cases}$$

i.e. it is band limited to $f = f_{max}$. Filtering out the main spectra would reconstruct the original spectrum without loss of information. We see immediately that the required condition is

$$f_{max} \leq \frac{1}{2\Delta t}$$

$$\Delta t \leq \frac{1}{2 f_{max}}$$

(10.3)

which is known as the Nyquist theorem, and

$$\frac{1}{2\Delta t} = \frac{1}{2} f_s$$

(where $f_s = 1/\Delta t$ is the sampling frequency) is known as the Nyquist frequency. For signal processing tasks, this defines the frequency range of the analysis. Often the sampling interval is normalized to 1, for sake of convenience. The normalized frequency range is then

$$\pm 0.5 \, [Hz]$$

A safety factor is used in practice, requiring a sampling frequency of 2.5 times f_{max}. In modern instrumentation systems, with binary divisions of frequencies generated by electronic clocks, a factor or $f_s = 2.56 f_{max}$ is very common.

Demonstrating the effect of incorrect sampling is relatively easy for the case of harmonic signals. Figure T10.4 shows the effect of sampling a harmonic signal of 10 [Hz] with a sampling frequency of

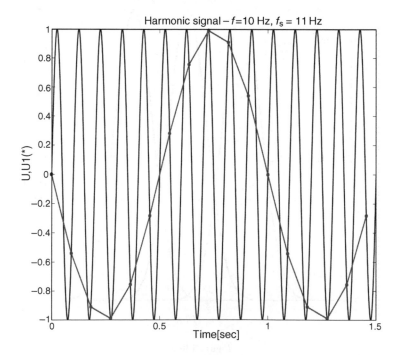

Figure T10.4

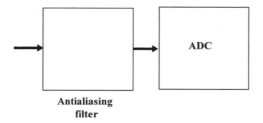

Figure T10.5

$f_s = 11$ [Hz]. Visual inspection shows that the samples would be indistinguishable from a harmonic signal of 1 [Hz] sampled at the same rate. The samples seem to describe a 1 Hz signal. This error is called an aliasing error. Computing a PSD would show power in the 1 Hz region! Aliasing results in high frequency components 'masquerading' as low frequency ones.

Formally we have an FT of the continuous signals as

$$\delta(f - 10) + \delta(f + 1) \tag{10.4a}$$

and for the sampled one (Equation 10.2)

$$\frac{1}{\Delta t} \sum_{n=-\infty}^{n=\infty} [\delta(f - 10 - nf_s) + \delta(f + 10 + nf_s)] \tag{10.4b}$$

And for $f_s = 11$, this results in frequencies of ± 1, 10, 12, 21, 23, 32. Limiting ourselves to the Nyquist frequency, this results in a component of 1 Hz, the difference between f and f_s.

In principle, an aliasing error cannot be corrected. It can only be avoided or minimized at the sampling phase. The original signal structure is destroyed by it, and it may not be meaningful to undertake any signal processing task on aliased data. Thus the aliasing error is usually the highest in the hierarchy of errors to be avoided.

10.4 Antialiasing Filters

Many analytic signals are not band limited. Actually a signal cannot be both time and band limited, as this would include all transients! In real life situations this would imply signals with a bandwidth beyond our required analysis range. There is a practical solution to avoid sampling errors when a signal has components which are beyond the Nyquist frequency, and that is to filter out any components beyond this range. Such a filter is called an antialiasing filter.

The antialiasing filter modifies the signal prior to the sampling operation. The aliasing error is thus avoided on the signal components passed by the filters. Obviously the signal components blocked do not appear in the analysis. The price paid for avoiding the error is that part of the signal must be ignored. In principle we need an analog filter to operate on the signal prior to the digitization (see Figure T10.5).

Specifications for antialiasing filters can be stringent. For example, we may require that the aliasing error at the Nyquist frequency be equal to the ADC resolution. Slopes of up to 120 dB/octave are in use. Analog realizations can be complex and expensive, and a simpler solution combining analog and digital filters can be used. This, however, is beyond the scope of this text.

11

Input/Output Identification

11.1 Objectives and Overview

Identification of a system is in a sense complementary to the analysis problem (Bendat and Piersol, 1993; Braun, 1986). In the analysis, a response is predicted based on knowledge of the system's model and the excitation. The identification extracts a model based on knowledge of an excitation and the response (Figure T11.1). It is usually based on experimentally acquired data, and measurement aspects have to be considered when performing such tasks. Actual identification methods thus utilize sampled data.

A major part of identification tasks assume that the system is linear, and thus draw heavily on the well-established discipline of linear system identification. As such we encounter cases of single input/single output (SISO), multiple input/single output (MISO) and multiple input/multiple output (MIMO). In this section it is mainly the SISO case which is addressed.

This chapter deals with direct identification in the frequency domain methods. The result of the identification will be the frequency response function (FRF), and as a possible by-product, its inverse transform, the impulse response. This identification is basically a 'nonparametric' one, as no physical or mathematical parameters describing the system are obtained. Unless some averaging is undertaken, the number of data points in the identified FRF equals those of the input or output data used in the identification task. Of course, system parameters can be extracted from the FRF at a later stage. However, there exist other, parametric methods, extracting parameters directly from the data (see Chapter 12).

Identification methods are also categorized according to the excitation signal used. The term 'rich' is used for excitations enabling us to identify more or less the complete system characteristics. In terms of the FRF, a rich excitation would cover the frequency range of interest. Rich excitations include transients, periodic and random functions. Harmonic excitations, usually called sine excitation, are very powerful, but not rich. A classification according to excitation methods thus covers stepped sine testing, impulse testing (transients), random testing, periodic burst, multi-sine, etc. The identification procedure and characteristics may be heavily dependent on the type of excitation used.

Utilizing experimentally acquired data, the various uncertainties associated with experiments necessitate a variety of approaches to the identification task. The majority of existing approaches address uncertainties which can be modeled as additive error, usually referred to as noise, to the measured signals. The assumptions made concerning such noise can dictate the type of identification procedure to be used. A specific procedure is geared to situations where uncertainties can be modeled as a white noise added to the response measurement only. Other procedures would handle noise added to the input or more general cases.

The accuracy of the results will depend greatly on the procedure, the correctness of assumptions and the type of excitation.

Discover Signal Processing: An Interactive Guide for Engineers   S. Braun
© 2008 John Wiley & Sons, Ltd

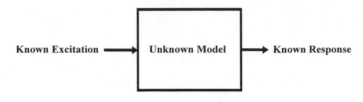

Figure T11.1

11.2 Frequency Domain Identification: The Noiseless Case

Frequency domain information is based on Fourier methods. Applying a Fourier transform to the differential equation describing the linear system results in an algebraic one.

We first start with a noiseless case: such a case is rarely assumed, but will be instructive. For our linear system (Figure T11.2)

$$Y(j\omega) = H(j\omega)X(j\omega)$$

$$H(j\omega) = \frac{Y(j\omega)}{X(j\omega)} \tag{11.1}$$

The FRF $H(j\omega)$ is a complex number. For the steady state response to a sine excitation of frequency ω, its magnitude is the gain at this frequency, and its phase the corresponding phase shift. Bode or Nyquist plots can be used to depict H graphically. The procedure would consist of computing the transforms of the excitations and responses and performing the division of Equation (11.1). The advantage of the frequency domain approach is evident from Equation (11.1): H is identified separately at each frequency. In principle an identification is possible, for limited frequency ranges.

Once the frequency response has been identified, it can be used to obtain the system's impulse response indirectly. This is based on the relation between the impulse response and the FRF via an inverse Fourier transform:

$$h(t) = \mathrm{F}^{-1}[H(j\omega)] \tag{11.2}$$

For the noiseless case, the identification would then depend on the signal class in the following manner:

(a) *Transient excitation*: Both the excitation and response would be broadband, covering a range of frequencies. A complete FRF can then be extracted from a single test. The identification can be performed as long as the value magnitude of $X(\omega)$ exceeds a minimum level. For too small $|X(\omega)|$ (and obviously for $|X(\omega)| = 0$) the problem is ill conditioned and the resultant H problematic. The frequency range of the identification is dictated by the richness of the excitation, i.e. by the region where excitation energy exists.

(b) *Sine excitation*: H is computed at a single frequency. Using a stepped sine, a function $H(\omega)$ would be computed step by step.

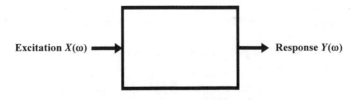

Figure T11.2

(c) *Random excitation*: This is usually broadband, i.e. rich, and a complete FRF can again be identified. Due to the statistical character of the excitation and response, the functions $X(\omega)$ and $Y(\omega)$ will have a probability distribution, which may be quite complex. The variance of spectral functions is significant, hence H as computed by Equation (11.1) may have a large variance as well. The variance can be reduced by averaging, but some care is needed in order to apply this correctly. Averaging of separately acquired functions $X(\omega)$ and $Y(\omega)$ would be incorrect, as their expectations $E[X(\omega)]$ and $E[Y(\omega)]$ are zero. Averaging (for the noiseless case) would be done as

$$\frac{1}{M}\sum_{i=1}^{M}\frac{Y_i(j\omega)}{X_i(j\omega)}$$

where X_i and Y_i are samples of computed frequency domain functions (Fourier transform), and M such functions are available, each acquired under the same stationary conditions.

11.3 Identification with Noise Corrupted Signals

The assumption is that excitation and/or response signals are corrupted by additive noise. It is realistic to assume that these additive noises are of random character.

11.3.1 Identification for Output Additive Noise (Figure T11.3)

This is probably the most popular model, even if the assumption that the noise can be modeled in such a way is not always justified.

$$Y(j\omega) = H(j\omega)\,X(j\omega) + N(j\omega) \tag{11.3}$$

Insight can be gained by first looking at a similar static problem, that of estimating a regression line (Figure T11.4)

$$y = ax + n$$

When the uncertainty n is zero, $a = y_i/x_i$, where any y_i or x_i (i.e single measurements) can be used. When n exist, a 'best' line is sought, with $\hat{a}$ an estimation of the slope. A least square criterion is used, resulting in

$$\hat{a} = \frac{\sum_i y_i x_i}{\sum_i x_i x_i} \tag{11.4}$$

The dynamic case Equation (11.3), is completely analog, and the LS solution is similar, except for the fact that H is complex.

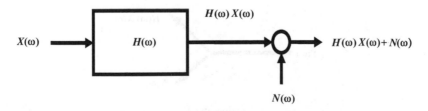

Figure T11.3

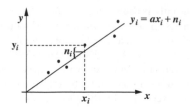

Figure T11.4

$$\hat{H}_1(j\omega) = \frac{\sum X^*(j\omega)Y(j\omega)}{\sum X^*(j\omega)X(j\omega)} = \frac{\sum S_{xy}}{\sum S_{xx}} = \frac{\tilde{S}_{xy}}{\tilde{S}_{xx}} \qquad (11.5)$$

In Chapter 7 we saw that the PSD of a random signal was computed by averaging estimators of the form $|X^2|$ (except for a constant multiplier, see Equation (7.5)). The denominator of Equation (11.5) is thus the PSD. For reasons to become obvious, it is called the autospectrum, denoted by S_{xx}, meaning 'x on x'. The numerator of Equation (11.5) is similarly called the cross-spectrum, denoted as S_{xy}. The Wiener–Khintchine: can be generalized to state that the cross-correlation R_{xy} and the cross-spectrum S_{xy} are related by a Fourier transform.

$$S_{xy}(\omega) \longleftrightarrow R_{xy}(\tau) \qquad (11.6a)$$

and we saw in Chapter 7 a special case with

$$S_{xx}(\omega) \longleftrightarrow R_{xx}(\tau) \qquad (11.6b)$$

The estimator (11.5) is often designated as H_1, and it is equal to the ratio of the cross-spectrum to the input's autospectrum.

An illuminating interpretation of the above identification process is possible. According to the model described by Equation (11.3), the identification of $H(j\omega)$ enables us to interpret $Y(j\omega)$ as comprising two parts (Figure T11.5):

- $H(j\omega)X(j\omega)$, that part of the response linearly related to the excitation
- an unrelated noise $N(j\omega)$.

In terms of signal terminology, $Y(j\omega)$ is considered to be composed of two orthogonal (independent) components. $H(j\omega)X(j\omega)$ is the coherent part of the response, $N(j\omega)$ the residual.

Having developed a method of identifying H, we now ask whether it is possible to have some confidence in the result. It is instructive again to look at the static analogy, shown in Figure T11.6 for two cases: while it is possible to compute a regression line in any case, Figure T11.6(b) is obviously meaningless. The almost intuitive solution would be to check whether the spread of the residuals around the estimated straight line is acceptable.

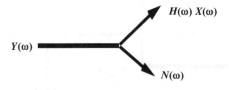

Figure T11.5

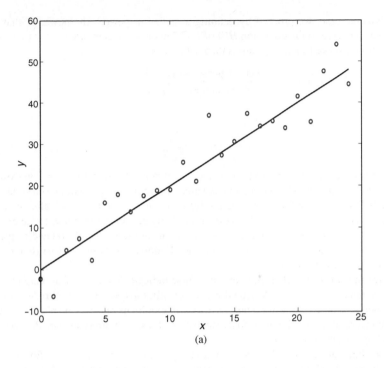

(a)

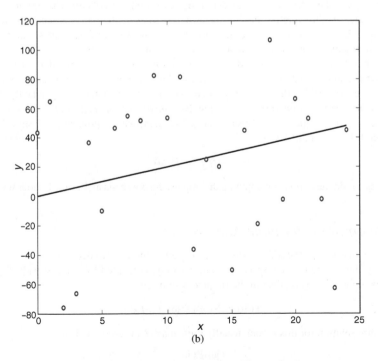

(b)

Figure T11.6

For the dynamic case, we proceed by defining a 'coherent power', the power of that part of the response linearly related to the excitation $H(j\omega)X(\omega)$. This is to be compared to the total power of the output $Y(\omega)$. A coherence function $\gamma^2(j\omega)$ is then defined as

$$\gamma^2(j\omega) = \frac{\text{Coherent response power}}{\text{Total output power}} = \frac{|S_{xy}|^2}{S_{xx}S_{yy}} \quad 0 \leq \gamma^2 \leq 1 \tag{11.7}$$

and estimated by

$$\hat{\gamma}^2(j\omega) = \frac{|\tilde{S}_{xy}|^2}{\tilde{S}_{xx}\tilde{S}_{yy}} \tag{11.7a}$$

The fact that the coherence function is both bounded and normalized to 1 makes it an extremely valuable criterion for the quality of identification. A value of 1 indicates an ideal noiseless case, where all the response is linearly related (via the system's dynamics) to the excitation. A value of zero indicates a meaningless result, when there is no linear relation of any part of the response to the excitation. The coherence function thus shows the degree to which the response is linearly related to the excitation. The existence of any noise in the measurements or of additional responses to other (linearly unrelated) excitations will decrease the coherence.

A coherence function of less than unity may of course indicate that the basic assumption, that of testing a linear system, is not correct. Any nonlinearity existing in the system will reduce the coherence function. It should be noted that while a nonlinearity will exhibit its presence by reducing the coherence function, the inverse does not follow: a coherence value of less than unity may be due to other causes, notwithstanding the perfectly linear system.

A coherence function of 1 will also result when a single pair of signals is used in Equation (11.7a). No residual can be computed for a least square estimate based on a single pair of observations, as will be obvious if analogy of the static case is considered. By definition, the coherence between any two purely signals would equal 1. This implies that it is meaningless to evaluate the coherence function for purely periodic excitation and responses. Power then exists only at discrete frequencies, and the coherence between two harmonic components of equal frequency must be equal to unity.

The coherence function can be considered as the frequency domain equivalent of the normalized input/output (time domain) cross-correlation function. The advantage of identification in the frequency domain is again noticeable: the coherence function can be evaluated independently for different frequencies. *The identification results can then be considered as meaningful for some frequency ranges, and unacceptable for others.* To summarize the identification procedure we show Figure T11.7. From Chapter 4, Equation (4.5) we had

$$h(t) \longleftrightarrow H(\omega)$$

Hence it is easy to add the option of computing the impulse response via an inverse Fourier transform of H.

11.3.2 Identification for Input Additive Noise

The assumption that most of the additive disturbances can be modeled as occurring at the excitation point may not always be justified. The response signal, x, can be corrupted by noise as well (Figure T11.8). Denoting this input noise by m, then in the frequency domain

$$Y(\omega) = H(\omega)[X(\omega) + M(\omega)] \tag{11.8}$$

The least square solution for this model, usually known as H_2, is computed from

$$H_2(j\omega) = \frac{\sum Y^*(j\omega)Y(j\omega)}{\sum X^*(j\omega)Y(j\omega)} = \frac{\sum S_{yy}}{\sum S_{xy}} = \frac{\hat{H}_1(j\omega)}{\gamma^2(j\omega)} \tag{11.9}$$

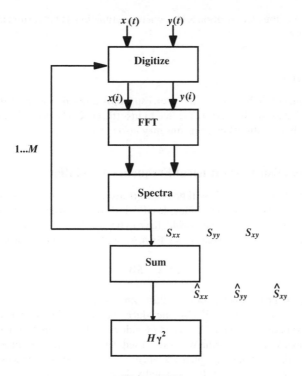

Figure T11.7

Both H_1 and H_2 can basically be computed from the same information and the general procedure of the figure. Some judgment is needed in order to decide which model, and hence estimate, to use. Each of the two estimators $\hat{H}_1$ and $\hat{H}_2$ correspond to different models of noise modeling. Bias errors in the identification will occur if the estimator used does not correspond to the actual noise situation existing.

11.4 Error Mechanisms and their Control in the Identification Process

The general types are similar to those occurring in spectral analysis. These are:

• Bias errors
• Random errors
• Leakage.

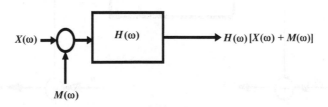

Figure T11.8

An essential quantifier of the identification is the coherence function. Hence errors have to be considered both for the FRF and γ^2.

11.4.1 Bias Errors

These are often considered as the most severe error, especially if resonance peaks in the FRF are under-estimated. In spectral analysis, bias errors are due to insufficient analysis resolution. The same occurs for the identification, but additional mechanisms may also occur.

11.4.1.1 Bias Errors Due to Insufficient Frequency Resolution

Most estimators computed via the DFT will be biased if insufficient frequency resolution is used. This resolution, Δf_1, must be small enough to follow local details of the estimated function. For resonating systems, local changes with resonance bandwidth have to be resolved. This bandwidth is approximately $2\zeta f_n = BW$ for small damping ratios (see Appendix 4.A), and the following is usually required:

$$\Delta f \leq \frac{1}{4} BW \tag{11.10}$$

and Δf can be chosen by the parameters of the Fourier analysis.

Insufficient frequency resolution will also introduce bias errors in the coherence function. The coherence will then be underestimated. The type of window used in the analysis has a significant effect on this underestimation. When a Hanning window is used, a too large Δf will cause the coherence function to exhibit sharp minima at those frequencies where H peaks, i.e. at resonances. With a rectangular window, the coherence function would show gradually decreasing values.

11.4.1.2 Bias Errors Due to Additive Noise Sources (Figure T11.9)

These errors are a function of the noise location as well as the estimator used. According to its definition/derivation, H_1 is insensitive to output noise, while H_2 is not affected by input noise. For all other cases, and with noise power spectral densities S_{nn} and S_{mm}, the following results are given for the estimators H_1 and H_2, where H_0 is the true FRF:

$$\left| \hat{H}_1 \right| = \left| H_0 \right| \left(1 + \frac{S_{mm}}{S_{xx}} \right)^{-1}$$

$$\left| \hat{H}_2 \right| = \left| H_0 \right| \left(1 + \frac{S_{nn}}{\left| H_0 \right|^2 S_{xx}} \right) \tag{11.11}$$

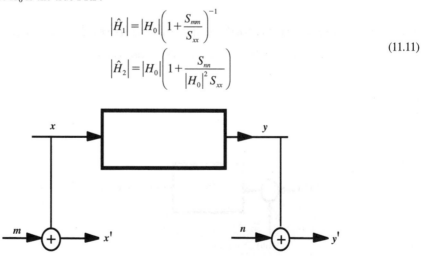

Figure T11.9

Figure T11.10

and

$$\left|\hat{H}_1\right| < \left|H_0\right| < \left|\hat{H}_2\right| \tag{11.11a}$$

These bias errors occur for the magnitude only. No bias error is introduced by the noises in the phase of the estimated FRFs.

The additive noises will obviously decrease the coherence function. It is easy to show that the computed coherence will be

$$\gamma^2 = \frac{1}{\left(1 + \dfrac{S_{mm}}{S_{xx}}\right)\left(1 + \dfrac{S_{nn}}{S_{yy}}\right)} \tag{11.12}$$

11.4.1.3 Bias Due to Unmonitored Additional Excitations

Usually it is assumed that the monitored excitation $x(t)$ is responsible for the monitored output $y(t)$. In actual tests, additional excitations may exist through unidentified paths (Figure T11.10). Should this additional excitation not be fully coherent with the monitored one, the output will not be fully coherent with the excitation, resulting in a coherence function less than unity. In addition the FRF will be biased. For example for H_1:

$$\hat{H}_1 = H_0\left(1 + \frac{Sx_1x_2}{Sx_2x_2}\right) \tag{11.13}$$

where x_1 is the monitored excitation and x_2 the unmonitored one. No bias error in $\hat{H}_1$ will occur if x_1 and x_2 are uncorrelated. $\hat{H}_2$ too will be biased, according to a slightly more complicated expression.

11.4.1.4 Bias Due to Delays Between Excitation and Response

A time delay can occur in many identification tasks. One classic example would occur in acoustic output measurements, where a propagation delay would exist for such a response. Delays can also occur due to linear phase shifts in some parts of the measurement system. A bias error will then be introduced in the identified coherence function. For random rich signals, it will be underestimated by a value dependent on the ratio of the time delay to that of the time duration of basic signal record (i.e. the one to be repeated for averaging purposes), as per Equation (11.14):

$$H_1 = H_0\left(1 - \frac{\tau}{T}\right) \tag{11.14}$$

11.4.2 Random Errors

These errors are due to the statistical variability of frequency domain estimators, when the excitations and responses are random, or include a random component. They do not occur with purely noise-free deterministic signals, either periodic or transient.

Similarly to autospectra estimates, those of FRF estimators based on purely random signals, not only exhibit a large statistical variance but are also inconsistent, i.e. this variance does not decrease with record length. Hence the need for the averaging process shown in Figure T11.7. Based on some simplifying assumptions, the normalized RMS error for the estimators has been developed (Bendat and Piersol, 1993) as follows:

$$e_{|H|} = \frac{\left[1 - \gamma_{xy}^2\right]^{1/2}}{|\gamma_{xy}|\sqrt{2M}} \tag{11.15}$$

As expected, this error is inversely proportional to the square root of M, the number of records used in the averaging process. It is also dependent on the coherence function, and very large errors occur when low coherence values, and hence no strong linear relationship, exist.

The use of Equation (11.15) is at best one of a guideline, as the coherence function itself can only be estimated by the data at hand. This estimate itself is prone to statistical errors, bias (see below) as well as the random one noted in Equation. It is usually the practice to apply windows to the data (see leakage below). This has the effect of increasing the correlation between adjacent spectral estimates. The number of degrees of freedom on which the FRF values are estimated increases, and hence the statistical variability decreases. A quantitative analysis of this effect can be found in (Bendat and Piersol, 1993).

11.4.3 Leakage Errors

This error, traceable to basic Fourier transform properties, occurs due to the induced periodicity of the signal spanned by the analyzed signal duration. It is usually effectively reduced by applying a window, which in this case has to be applied to both the excitation and responses. A Hanning window is often the preferred choice for random signals. As mentioned above, this also has the extra beneficial result of reducing random errors.

Similarly, a rectangular window, spanning only the excitation duration (thus giving zero weight outside this span), could be used in order to enhance the effective S/N at the input.

11.5 Estimation Errors for the Coherence Function

Bias and random errors will exist for the estimator computed by Equation (11.7a). Bias errors are usually due to insufficient computational resolution Δf. The type of error depends, however, on the window used (Braun, 1986):

- For the rectangular window the error is one of underestimation.
- For a Hanning window, the underestimation is also aggravated by a phase gradient of the transfer function H. For example, around a resonance, where the phase gradient is maximized, a too small Δf can cause a dip in the coherence function.

Similar to Equation (11.15), random errors for γ^2 can be estimated by the expression

$$e_{|\gamma^2|} = \sqrt{2}\frac{\left[1 - \gamma_{xy}^2\right]}{|\gamma_{xy}|\sqrt{M}} \tag{11.16}$$

And again very large errors occur when low coherence values, and hence no strong linear relationship, exist.

An interesting case occurs when $M = 1$, when the coherence between a single realization of input/output signals is estimated. The LHS of Equation (11.7a) then always equals unity, a perfect coherence at all frequencies! An easy interpretation of this case follows from Figure T11.4, the equivalent case being the 'best line' passing through a single measured x, y value.

12

Model-based Signal Processing

12.1 General

Many signal processing techniques, relatively modern ones, extract models from measured data (Candy, 2005; Hayes, 1996). To introduce the concept, consider a signal $x(i\Delta t)$, with $i = 0, 1 \ldots N$. It may be possible to model the signal with a number of parameters p, where $p < N$. Signal models can thus be used for an efficient signal representation and are extensively applied in areas of diagnostics/ classification, spectral analysis and data compression, classification, diagnostics and spectral analysis techniques. The methods described below are sometimes denoted as 'parametric', whereas the Fourier-based methods, where an N data sequence in one domain is transformed to N data points in another, are nonparametric.

12.2 Signal Models

12.2.1 Stochastic Models

These signals include random process models, time series models that approximate discrete time processes encountered in practice, specifically sampled signals acquired from real systems. Three major types, AR, MA and ARMA, are based on the rational polynomial form in the Z-domain, as briefly introduced in Chapter 4, Equation (4.8).

One simple interpretation is to model these processes as being generated by filtering white noise by linear shift invariant filters.

MA (Moving Average) Process

$$x_i = w_i + b_1 w_{i-1} + b_q w_{iq} + \ldots b_q w_{i-q} = \sum_{k=0}^{q} b_1 w_{i-k}$$

In matrix notation: $\hspace{8cm}$ (12.1a)

$$x_i = \mathbf{w}^r \mathbf{b}$$
$$\mathbf{w} = \begin{bmatrix} w_i & w_{i-1} \ldots w_{i-q} \end{bmatrix} \qquad \mathbf{b} = \begin{bmatrix} 1 & b_1 \ldots b_q \end{bmatrix}$$

In the context of linear systems, w is the system's input (white noise) and x the output. Applying a Z transform,

Discover Signal Processing: An Interactive Guide for Engineers   S. Braun
© 2008 John Wiley & Sons, Ltd

$$X(z) = \sum_{k=0}^{q} b_k z^{-k} W(z) = B(:z)W(z) \tag{12.1b}$$

with $B(z)$ a rational system, an all-zero system with q zeros.

AR (autoregressive) Process

$$x_i = -a_1 x_{i-2} - a_2 x_{i-1} - \ldots a_1 x_{i-p} + w_i = -\sum_{k=1}^{p} a_k x_{i-k} + w_i \tag{12.2a}$$

The name autoregressive indicates that this is a linear regression of x on itself, with w the residual:

$$X(z) = \frac{1}{1 + \sum_{k=1}^{p} a_k z^{-k}} W(z) = \frac{1}{A(z)} W(z) \tag{12.2b}$$

with $1/A(z)$ an all-pole system with p poles.

ARMA (Autoregressive Moving Average) Process

$$x_i = -\sum_{k=1}^{p} a_k x_{i-k} + \sum_{k=0}^{q} b_k w_{i-k} \tag{12.3a}$$

$$X(z) = \frac{B(z)}{A(z)} w(z) \tag{12.3b}$$

with $B(z)/A(z)$ a rational system with p poles and q zeros.

12.2.2 Deterministic Models

These are based on equating a signal to the impulse response of a linear shift invariant filter, having a rational system function (Figure T12.1):

$$X(z) = H(z) = \frac{B(z)}{A(z)} \tag{12.4a}$$

where $B(z)$ and $A(z)$ have q and p zeros respectively. The matching of a given signal x_i to such an impulse response is called a Padé matching.

This time domain formulation can be formulated as

$$x_i = \sum_{k=1}^{p} c_k r_k^i \tag{12.4b}$$

where r_k are the roots of the denominator of $A(z)$ and c_k are the complex coefficients of the expansion. This is also known as a Prony model. The roots of r_k may be real, but often (for example for vibration signals) complex, describing decaying oscillatory signals:

$$x_i = \sum_{k=1}^{p} A_k \exp(\alpha_k + j2\pi fk) \tag{12.4c}$$

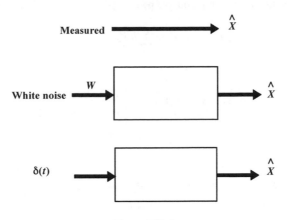

Figure T12.1

12.3 Modeling of Signals

The signal models are used to approximate data. Thus the data generated by the model will approximate measured data, and the type of the approximation needs to be understood. For spectral analysis we will require that the autocorrelation (and hence the PSD) of the data generated by the model corresponds to that of the observed data (Figure T12.1)

The modeling task consists of two major phases. In the first, the model's structure is defined, say an AR or ARMA one. This includes the determination of the model's order, say p in the AR case. The second phase is that in which the model's parameters are found, such that the 'best' approximation results. A criterion has to be defined and used. The parameters themselves may have a direct physical interpretation. A synthetic model, where the parameters are purely mathematical, can still be used for the purposes of simulation, characterization, etc.

12.3.1 Modeling of Stochastic Signals

Some of the most basic approaches are described below. Commercial software is now available, enabling us to apply a multitude of modeling techniques, and only an introductory presentation is given in this text.

The concept of the correlation matrix, encountered below, is briefly given in Appendix 12.A.

ARMA Models

From Equation (12.3c), multiplying by x_{i-k},

$$R_x(1) = -\sum_{k=0}^{p} a_k(R_x(1-k)) + \sum_{k=0}^{q} b_k R_{wx}(1-k) \tag{12.5}$$

where R are correlation functions. If we consider x_i as the result of filtering the white noise w by a filter with impulse response h,

$$x_i = h_i \otimes w_i$$

then

$$R_x(1) + \sum_{k}^{p} a_k R_x(1-k) = \sigma_w^2 \sum_{k=0}^{q} b_q h_{k-1} = \sigma_w^2 c_k \tag{12.6}$$

where we have denoted the term multiplying σ_w^2 by c_k

$$c_k = \sum_{k=1}^{q} b_k h_{k-1}$$

Thus

$$R_x(1) + \sum_{k=1}^{p} a_k R_x(1-k) = \begin{cases} \sigma_w^2 c_k & 0 \leq 1 \leq q \\ 0 & 1 > q \end{cases}$$ (12.7a)

which for $k = 1, 2$

$$\begin{bmatrix} R_x(0) & R_x(-1) & \cdots & R_x(-p) \\ R_x(1) & R_x(0) & & R_x(-p+1) \\ \vdots & & & \\ R_x(q) & R_x(q-1) & & R_x(q-p) \\ \cdots & \cdots & \cdots & \cdots \\ R_x(q+1) & & & R_x(q-p+1) \\ R_x(q+p) & & & R_x(q) \end{bmatrix} \begin{bmatrix} 1 \\ a_1 \\ \vdots \\ a_p \end{bmatrix} = \sigma_w^2 \begin{bmatrix} c_1 \\ \cdot \\ \cdot \\ \cdot \\ c_q \\ \cdots \\ 0 \\ \cdot \\ 0 \end{bmatrix}$$ (12.7b)

which are known as the *Yule Walker* (YW) equations.

AR Models

AR models are probably the most widely used ones. One of the reasons is that the necessary computations are based on the solution of a set of linear equations. From Equation (12.5) with $q = 0$ we have

$$R_x(1) = -\sum_{k=1}^{p} a_k R_x(1-k)$$ (12.8a)

Using $k = 1, 2...$we get

$$\begin{bmatrix} R_x(0) & R_x(+1)... R_x(-p) \\ R_x(1) & R_x(0) \\ \vdots \\ R_x(p) & R_x(0) \end{bmatrix} \begin{bmatrix} 1 \\ a_2 \\ a_2 \\ \vdots \\ a_p \end{bmatrix} = \sigma_w^2 \begin{bmatrix} 1 \\ 0 \\ \vdots \\ 0 \end{bmatrix}$$ (12.8b)

Or in matrix notation

$$\mathbf{Ra} = \sigma_w^2 \mathbf{I} \quad \text{with} \quad \mathbf{I} = [1\,0\,0\,0]$$

and the modeling consists of computing the elements a of $\mathbf{a}$. An estimate of $\mathbf{a}$ can be solved by estimating $R(k)$ based on available data, i.e. by

$$R(1) = E[x_k\, x_{k-1}]$$ (12.8c)

Some available procedures are based on a slightly different interpretation of the AR model Equation (12.3a). This can be considered as linear prediction, where x_i is a linear combination of past observations x_{i-k}, and w is the residual or prediction error. This is a forward prediction procedure. A backward prediction can also be defined, where it is desired to predict the 'next earlier' value as per

$$x_{i-p} = -b_1 x_{i-p+1} - b_2 x_{i-p+2} + \ldots\ldots w_{bi} \tag{12.9}$$

It can be shown that the YW equation

$$\mathbf{Rb} = \sigma_{w_b}^2 \mathbf{I} \tag{12.10}$$

can be used to solve $\mathbf{b}$ which minimizes the mean backward error w_b (in some specific sense).

12.3.2 Least Square Estimation Methods for AR Modeling

These are based on casting the AR prediction equations as a set of linear equations in the parameter vector $\mathbf{a}$, and minimizing the mean square of the prediction error. From Equation (12.3)

$$w_i = x_i + \sum_{k=1}^{p} a_k x_{i-k} \tag{12.11}$$

A matrix equation can be cast as

$$
\begin{bmatrix} w_1 \\ w_p \\ w_{pn} \\ \cdots \end{bmatrix}
=
\begin{bmatrix}
x_0 & 0 \\
\vdots & \vdots \\
\vdots & \vdots \\
x_{p-1} & x_0 \\
x_p & x_1 \\
\vdots & \vdots \\
x_{n+2} & x_{w-p+1} \\
0 & x_{w-p+1} \\
0 & x_{w-1}
\end{bmatrix}
\begin{bmatrix} a_1 \\ \vdots \\ a_p \end{bmatrix}
=
\begin{bmatrix}
x_1 \\
\vdots \\
\vdots \\
x_p \\
x_{p+1} \\
\vdots \\
x_w \\
x_{n+1} \\
\vdots \\
x_{n+1}
\end{bmatrix}
\tag{12.12a}
$$

$$\mathbf{w} = \mathbf{X}\,\mathbf{a} + x \tag{12.12b}$$

with the LS solution

$$\hat{\mathbf{a}} = (\mathbf{X}^T\mathbf{X})^{-1}\,\mathbf{X}^T\mathbf{x} \tag{12.13}$$

The expression $\mathbf{X}^T\mathbf{X}$ is a correlation matrix, based on observed data, encountered already in Equation (12.13). Correlation functions computed from data are used. In that sense it is similar to the solution of the YW equations, since computed estimators of correlation functions, and not the unknown theoretical correlation functions, are the only ones available for the solution.

12.3.3 Sequential Methods

A time-varying modeling is possible by a sequential, as opposed to a block, analysis. Techniques developed in the areas of adaptive filters can be used. As an example let us mention the least mean square (LMS) method, where coefficients are updated as new data is incoming. This can be useful to track signals with relatively slow time-varying characteristics.

For the AR polynomial, the coefficients a_k will now be time dependent. Thus Equation (12.11) will become

$$w_i = x_i + \sum_{k=1}^{p} a_{i-1,k} x_{i-k}$$ (12.14)

where the index 1 denotes the time dependency of **a**. The coefficient vector is now updated as per

$$\mathbf{a}_i = \mathbf{a}_{i-1} + \Delta \mathbf{a}_{i-1}$$ (12.15a)

and the following updating is the so-called LMS algorithm:

$$\mathbf{a}_i = \mathbf{a}_{i-1} - \mu \nabla \mathbf{x}_{i-1}$$ (12.15b)

μ is the adaptive time constant which affects the adaptive performance, including its convergence.

Many additional adaptive schemes exist and are available, all beyond the scope of this text. Sequential methods are the obvious choice for the analysis of nonstationary signals.

12.3.4 Parametric Methods for Deterministic Signals

Model-based methods can also be applied for cases involving deterministic signals like decaying transients, harmonic signals, etc. One approach is to approximate a signal x by an impulse response h of a linear shift time invariant filter. For the general case, the transfer function of this filter is as in Equation (12.4a):

$$H(z) = \frac{B(z)}{A(z)}$$

and the coefficients of H, a_k and b_k of A and B are found by minimizing the error term:

Minimize
$$E = \sum |e|^2$$ (12.16a)
$$e = x - h$$

$$\frac{\partial E}{\partial a_k} = 0 \quad k = 1, 2 ... p$$
(12.16b)
$$\frac{\partial E}{\partial b_k} = 0 \quad k = 1, 2 ... q$$

results in a set of equations which are nonlinear in a and b. Direct methods of solving are thus of iterative form like the method of steepest descent and others.

Other methods solve separately for two sets of parameters via linear equations. The so-called Prony method is one of that class. Here an overdetermined set of linear equations solves $A(z)$ and $B(z)$ of Equation (12.4a). The Prony method has a basis function of frequencies directly determined from the data, computed from the zeros of the polynomial $A(z)$.

Available Algorithms

Various computational schemes are available to the analyst, some of them dependent on the range used. The method utilizing the range $1 \cdots n + p$ is called the autocorrelation method. It computes by 'forcing' x to be zero outside the observed interval, i.e. windowing. The Toeplitz structure is, however, retained, enabling the use of the Levinson recursion. Utilizing the range $i = p + 1 \ldots$ is called the covariance method. No windowing is now implied, but the Toeplitz structure no longer exists. It can be shown that the autocorrelation method is actually equivalent to the Yule-Walker method.

A modified covariance method is based on a similar approach, but minimizes the forward and backward error terms. The names are misnomers, for historical reasons, and have no relation to the statistical functions of autocorrelation and covariance.

Another procedure, Burg's method, also minimizes the sum of squares of the forward and backward prediction errors, but uses a recursive procedure up to the pth parameter a_p, with the advantage that the resulting $A(z)$ is always stable.

Most of the above procedures, including Prony modeling, are well documented and readily available in commercial as well as public domain software packages.

12.3.5 Model Order and Overdetermination

Experimentally acquired signals usually include an additive noise component. The accuracy of the parameters computed by model-based methods are highly affected by the existing signal to noise ratio and the model order chosen. Choosing too large a model order will result in poles which model noise terms. It is sometimes possible to estimate the signal model order via the rank of the covariance matrix of the signal.

An interesting phenomenon occurs, however, with overdetermination of p, where p is the correct model order. In addition to generating noise-related poles, it is found that the signal-related poles will get closer to the true ones in the presence of noise. Thus it can be beneficial to choose too large a model order. The correct signal poles can be recognized by consecutively increasing the model order. A clustering of identified parameters will occur for the correct ones, while the location of the noise-related ones will be all over the parameter space. This type of test is sometimes called a stabilization check.

12.4 Model-based Spectral Analysis (Stoica and Moses, 2002)

Some advantages are claimed for this approach, when compared to Fourier-based PSDs. One of them is an improved performance in resolving components of close frequencies in cases of limited, short data sequences.

12.4.1 Procedure

Once a rational signal model is available, then the computation of the PSD is basically straightforward. For the linear system,

$$S_x(\omega) = S_w(\omega)|H(\omega)|^2 \tag{12.17}$$

where S_x is the PSD of the signal, and S_w that of the innovation process (or residual). H is the FRF of the signal model. With

$$S_w(\omega) = S_w = \text{con} = P_w \, \Delta T \tag{12.18a}$$

$$S_x(\omega) = P_w \Delta t \left|\frac{B(\omega)}{A(\omega)}\right|^2$$

where

$$B(\omega) = B(z)\big|_{z=\exp(i\omega\Delta t)}$$
$$A(\omega) = A(z)\big|_{z=\exp(i\omega\Delta t)} \tag{12.18b}$$

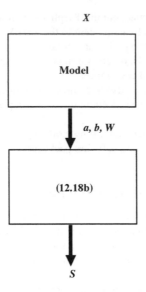

Figure T12.2

The PSD estimation thus consists of three basic steps (Figure T12.2):

- Step 1: Selection of the signal model, including an appropriate order
- Step 2: Estimation a of the model parameters
- Step 3: Insertion of the parameters into the theoretical PSD (Equation (12.18)).

The PSD computed is based on model parameters, hence is a parametric method. This is to be compared to the FFT-based analysis, a nonparametric method, where an N point sequence in the time domain is transformed into an N point sequence in the frequency domain except for additional points used for segment averaging (see Chapter 7).

The model-based PSD is much smoother than the FFT-based one, as it is described by a small number of parameters (see Figure T12.3). In this sense we have a type of data reduction, describing our data by a smaller number of parameters.

The statistical properties of the estimated PSD depend heavily on the number of data points versus the number of model parameters. For the AR case, the result is asymptotically unbiased for large N. As parametric methods are often used in view of their improved frequency resolution, i.e. for time-limited records, such knowledge is of limited value. The variance of the estimators are roughly proportional to p/N, with p the model order.

12.5 Model or Selection

The smoothness of the PSDS function as well as the frequency-resolving capabilities, are all strongly dependent on the chosen model order. One convenient point of view is that of data reduction, when $p + q$ parameters (plus the innovation process power) describe the characteristics of the data.

For ARMA models, a search for reasonable orders is often an option, where 'reasonable' is determined by heuristic arguments. Objective model orders, based on the minimization of some specific criterion, exist mainly for AR models.

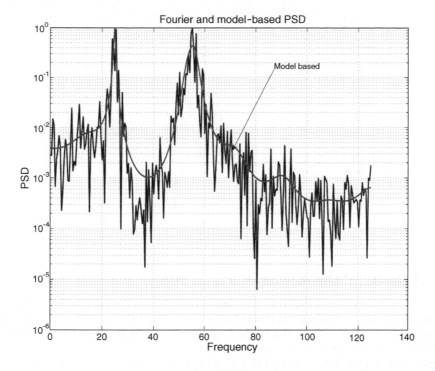

Figure T12.3

The following criteria achieve some balance between a high model order which causes a decrease in the prediction error power, i.e. a penalty increasing with the model order:

- The final prediction error (FPE) criterion:

$$\text{FPE}(p) = \sigma_w^2 \frac{N+(p+1)}{N-(p-1)} \tag{12.19}$$

A smaller p/N ratio uses, on average, fewer data points per parameter a_k of the AR model. This results in a less (statistically) accurate parameter. The order p minimizing $\text{FPE}(p)$ is then used as the optimal one.

- Another is the Akaike information criterion (AC):

$$\text{AIC}(p) = N\ln(\sigma_w^2) + 2p \tag{12.20}$$

An information theoretic function is minimized. Actually $\text{AIC}(p)$ and $\text{FPE}(p)$ will be asymptotically equivalent as N increases.

The practical choice of p is often problematic, as no clear minimum of any criterion may be evident. A local minimum may be erroneously chosen. Often the region where a criterion 'plateaus' (flattens out) is the best alternative.

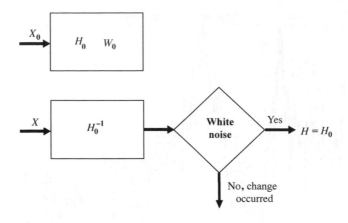

Figure T12.4

12.6 Model-based Diagnostics (Wu *et al.*, 1980)

When a correct signal model is available, the suitability of this model to describe other signals can be tested. A test can be based on the properties of the residual (sometimes called innovation) sequence, which should be a white noise when the model is correctly identified for a specific model. This is the basis for some model-based diagnostic methods.

The approach is described by Figure T12.4. A model is first identified from a test signal. With the information from this signal and the identified model parameters, the residual sequence can then be computed via Equation (12.11):

$$w_i = x_i + \sum_{k=1}^{p} a_k x_{i-k}$$

Computing this residual for a new signal x can help us check the hypothesis that the model describing the original and new signals is the same: if $H = H_0$, e has white noise properties, else $H = H_1$, i.e. a change occurred. The white noise properties can be checked by various methods. One possibility is to compute the autocorrelation, which should be a Dirac impulse function (see Chapter 7).

It is worthwhile noting that a change can be detected, without indication of its cause. If continuous monitoring is performed via a recursive updating of the identified model parameters, then the time when changes occur can be detected.

Appendix 12.A The Correlation Matrix

For the discrete case we have

$$R_{xy}(k) = \frac{1}{N - |k|} \sum_{n=0}^{N-1-|k|} x(n) y(n + |k|)$$

for the unbiased one, and   (12.A1)

$$R_{xy}(k) = \frac{1}{N} \sum_{n=0}^{N-1} x(n) y(n + |k|)$$

The products of correlation values (for the range of delay indexes) can be arranged in a matrix form known as the correlation matrix. Many digital modeling and filtering operations involve the use of such matrices.

For discrete data, the autocovariance and autocorrelation sequences are important second-order characterizations. For the discrete sequence

$$\mathbf{x} = [x(0)x(1) \dots x(p)]^T$$

These are defined via the $(p+1)x(p+1)$ outer product

$$\mathbf{x}\mathbf{x}^T = \begin{bmatrix} x(0)x(0) & x(0)x(1) & \dots & x(0) & x(p) \\ x(1)x(0) & x(1)x(1) & \dots & x(1) & x(p) \\ x(p)x(0) & x(p)x(1) & \dots & x(p) & x(p) \end{bmatrix} \tag{12.A2}$$

For a wide sense stationary process, the autocorrelation R_x matrix is based on the expectation of Equation (12.A2):

$$\mathbf{R}_x = \mathrm{E}(\mathbf{x}\mathbf{x}^T) = \begin{bmatrix} r_x(0) & r_x(1) & & r_x(1) & r_x(p) \\ r_x(1) & r_x(0) & & r_x(1) & r_x(p-1) \\ \vdots & & & & \\ r_x(p) & r_x(p-1) & & & r_x(0) \end{bmatrix} \tag{12.A3}$$

13

Machinery Diagnostics: Bearings and Gears

13.1 Diagnostics and Rotating Machinery

Vibration-based diagnostics is often applied to rotating machinery (Braun, 1986). Specific signal characteristics can be correlated to specific machine elements. The term 'signature' is often used to describe signals measured on machines and their elements, which can be indicative of their mechanical integrity.

Models describing the signal generation process are very important in order to extract relevant information from these signatures. For rotating machinery and their elements, these are usually based on the relevant geometries and kinematics. Very often the signals used as information carriers for diagnostic purposes are vibration signals, body or airborne. Decomposing such signals can reveal components with frequencies which are tracking the basic rotational speed, and hence carry information concerning the status of rotors, bearings, gears, couplings, blades, etc. The transmission path from the signal generation source to the monitoring location is usually very complex. It may involve multiple propagation paths, frequency-dependent speed of propagation and attenuation. Relating other signal characteristics (shape, magnitude, etc.) to mechanical integrity is often much more difficult, and frequency (or period) analysis is thus prevalent.

This chapter will deal briefly with rotor imbalance, and then with the signal generation model for two types of elements: roller bearings and gears.

13.2 Structural Effects

The complex transmission path from the excitation source to the measurement location manifests itself by a complex frequency response function (FRF) between them. A significantly modified signal is thus measured as a vibration response. It is worthwhile mentioning that the dynamics of the measuring device (see Section 4.3) must be considered a part of the general modifying system. Thus

$$X(f) = H_{\text{stru}}(f)X_{\text{ex}}(F)$$

where X, X_{ex} and H_{struc} are the frequency domain representation of $x(t)$, $x_{\text{ex}}(t)$ and $h_{\text{struc}}(t)$, the measured response, excitation and the structure's impulse response. H_{struc} usually shows many resonance regions, with high magnification factors due to low damping ratios. Any excitation whose frequency range lies within a resonance region will thus be highly amplified.

Discover Signal Processing: An Interactive Guide for Engineers   S. Braun
© 2008 John Wiley & Sons, Ltd

In real life situations, multiple excitations exist, as any machine will have multiple components, each one generating exciting forces. When attempting to analyze a component of $x(t)$ which is indicative of a specific mechanical component fault, it is often the practice to filter $x(t)$, so as to analyze regions where the component of interest is of high energy, as compared to all others. The filtering process attempts to improve the signal to noise ratio for the desired signal.

13.3 Rotating Imbalance

For machines with rotors rotating around a fixed axis, irregularities in the mass distribution result in a harmonic radial force, with a frequency equal to the rotating frequency. This force is minimized by mass balancing, where mass removal (or addition) results in a canceling force component. The vibration resulting from mass imbalance thus has the form

$$x(t) = A\sin(2\pi f_r t + \phi) \tag{13.1}$$

with f_r the rotating speed (in Hz) and ϕ a function of the location of the equivalent mass imbalance to a defined reference point. For machines rotating at variable speeds (for example during start-up or shutting down), f_r will be a function of time.

13.4 Modeling of Roller Bearing Vibration Signals (McFadden and Smith, 1984; Oehlmann *et al.*, 1997)

The basic measurement scheme is shown in Figure T13.1.

Deterioration of roller bearings usually begins via a localized defect on the inner or outer race. An impacting shock is produced as each rolling element (ball or cylinder) passes the defect (Figure T13.2).

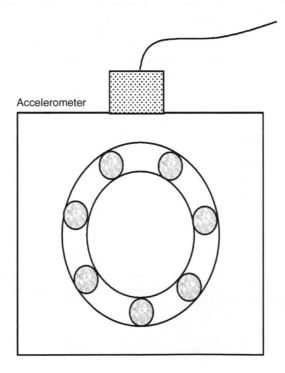

Figure T13.1

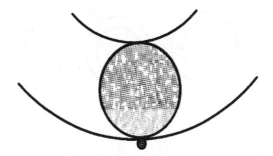

Figure T13.2

The measured vibrations are then the response to this shock as it propagates to the external monitored location.

Denoting the single shock as $x_{0\_ex}(t)$, the signal is periodic with $x_{ex}(t) = x_{0\_ex}(t + T_{sh})$.

$$x_{ex}(t) = \sum_r x_{0\_ex}(t - rT_{sh}) \tag{13.2}$$

From geometrical and kinematic considerations, it can be shown that the frequency $1/T_{sh}$ depends on whether the localized defect is on the outer (f_o) or inner race (f_i). For a fixed outer race and rotating inner race, and for bearings designed for radial loading, these are given by

$$f_o = \frac{n}{2}f_r\left(1 - \frac{d}{D}\right)$$

$$\tag{13.3}$$

$$f_i = \frac{n}{2}f_r\left(1 + \frac{d}{D}\right)$$

where f_r is the rotating frequency of the inner race, n the number of rolling elements, d the diameter of the rolling elements and D the diameter from the center of the rolling elements. For bearings suited for axial loading, the force applied from the rolling elements to the races is at an angle called the 'contact angle', and the force component applied between the rolling element to the races occur at a 'contact' angle α, and frequencies predicted by Equation (13.3) will be slightly modified.

Figure T13.3 shows a signal corresponding to the case of a localized defect on the outer race. For such periodic signals, the spectrum is obviously composed of the spectral lines at

Outer race fault frequency components: kf_o, $k = 1, 2, 3...$ $\tag{13.4}$

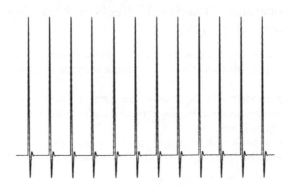

Figure T13.3

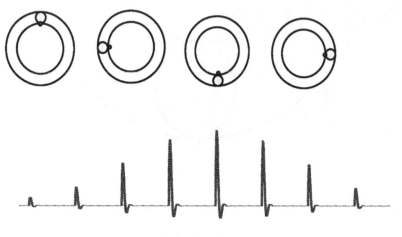

Figure T13.4

In the case of a localized defect on the inner race, any external radial load (for example that due to the rotor weight for a horizontal rotor) will affect the resultant signal. This is then typically of the form shown in Figure T13.4. where the small circle shows the temporary location of the defect. The signal is modulated by the rotation frequency. It is of maximum amplitude when the impulsive force due to the ball passing on the defect is aligned with the external radial load, and minimum 180 degrees later. The shock excitation sh(t) thus has the form

$$\left[1 + g(t)\right] \text{sh}(t) \tag{13.5}$$

where $g(t)$ is the modulation signal (see par 3.11). Its shape is affected by the loading zone function, resulting from the applied radial load. It is a periodic function, of frequency f_r, with each period in the form of uni-mode window peaking at the center. When approximating $g(t)$ as the harmonic function, the shock excitation spectrum will be a line spectrum with frequencies of

$$kf_i \pm f_r$$

or for the more general case, where $g(t)$ is periodic but not necessarily harmonic (and hence has frequencies qf_r, q = 1,2,...)

Inner race fault frequency components: $kf_i \pm qf_r$ $k = 1, 2, 3...$ $q = 1, 2, 3...$ \qquad\qquad (13.6)

13.5 Bearing Vibrations: Structural Effects and Envelopes
(Prashad *et al.*, 1985; White, 1991)

Localized faults generate sharp impulsive excitations, whose frequency energy distribution is thus wideband. The response spectrum then exhibits the multiple resonance regions of the structural system (Figure T13.5).

It is often of advantage to analyze the envelope of such a signal. This has the form of a low frequency signal, enveloping the oscillations inside the envelope, and can be considered to approximate the original excitation shocks (Figure T13.6).

Let us assume that the response to the excitation shocks are approximated by the impulse response of the structural system, $h(t)$. Then from Equation (13.2), assuming an inner race fault (frequency f_i),

$$x(t) = \sum_r Ah\left(t - \frac{r}{f_i}\right)$$

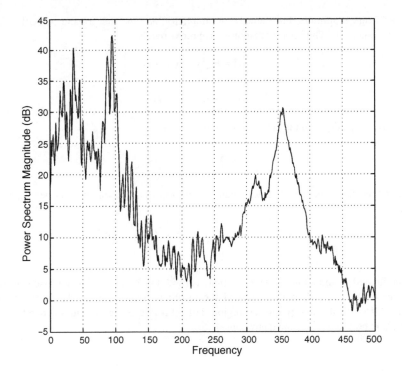

Figure T13.5

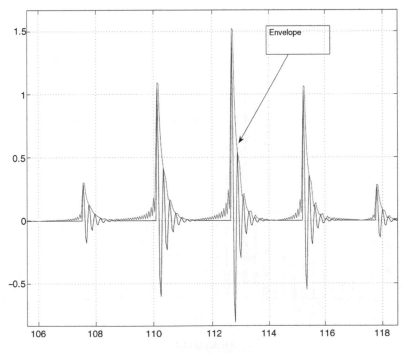

Figure T13.6

with A an arbitrary amplitude. For the sake of simplicity, assume that a band pass filter is applied, around one resonance f_0. The impulse response is approximately

$$h(t) = h_{env}(t)\sin(2\pi f_0 t)$$

$$h_{env}(t) = \exp(-2\pi\varsigma f_0 t)$$

Applying an envelope detection to $x(t)$ will result in

$$x_{env} = \sum_r Ah_{env}\left(t - \frac{r}{f_i}\right)$$

For an inner race fault, any radial load generates an amplitude modulation as per Equation (13.5) resulting in

$$x_{env}(t) = \left(1 + \sum_r A\left[1 + g\left(t - \frac{r}{f_i}\right)h_{env}\left(t - \frac{r}{f_i}\right)\right]\right)$$

This is a modulated periodic low frequency signal, as shown in Figure T13.6 (see also Chapter 8).

The power of $X(f)$ is concentrated around the resonant frequencies of the system. That of the envelope is shifted to the low frequency region, but retaining the pertinent sideband patterns (see Figure T13.7, a and b spectra, c and d envelopes).

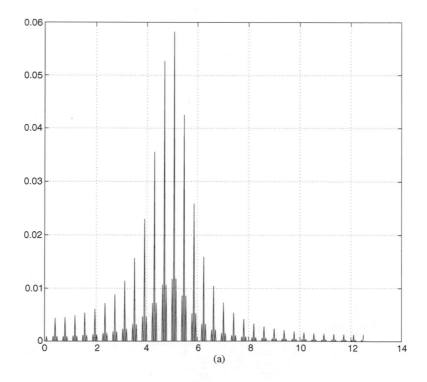

(a)

Figure T13.7

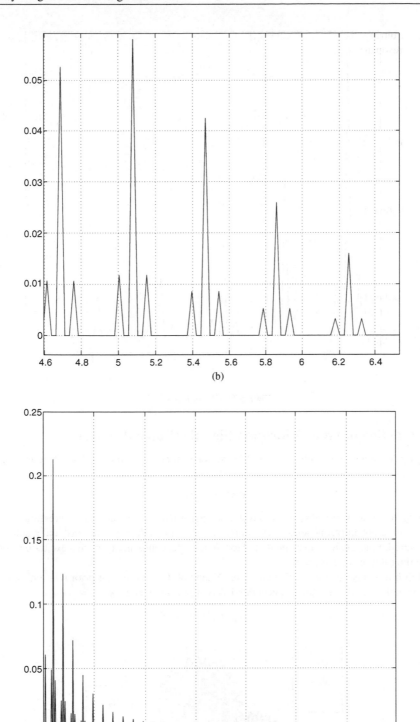

Figure T13.7 (*continued*)

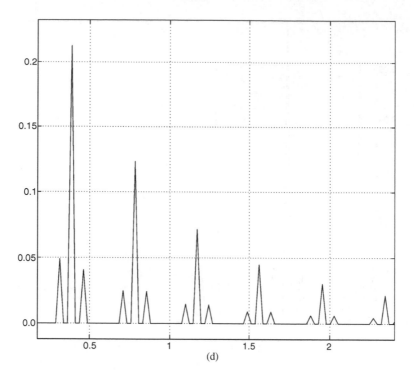

(d)

Figure T13.7 (*continued*)

13.6 Modeling of Gear Vibration Signals (Randall, 1982)

Denoting by N the rotating frequencies, and M the corresponding number of teeth (Figure T13.8), the basic relation for the gear pair is

$$N_1 M_1 = N_2 M_2 \tag{13.8}$$

Engineering units for N are often in rpm (rotations per minute), and then the expressions have to be divided by 60 to obtain frequencies in Hz. Forces are generated during rotations, and resultant vibrations measured. Frequencies can be predicted for some of the components of such excitation signals, all related to the rotational frequencies.

In addition to the basic rotational frequencies N_1 and N_2 (related by the prior relation), there is the component at the *meshing* frequency, generated as two gear teeth engage

$$f_m = N_1 M_1 = N_2 M_2 \tag{13.9a}$$

Figure T13.8

Meshing frequency harmonics: kf_m, $k = 1, 2, 3...$ (13.9b)

Modulation effects can occur. Due to eccentricities (when the geometric and rotational center do not coincide exactly), amplitude modulation once per rotation can occur. Speed fluctuations can also occur in such cases (as the load on the driving system can affect the rotational speed), resulting in frequency modulation, again at once per rev. Hence both amplitude and frequency modulation can occur simultaneously. For such a periodic (but not necessarily harmonic) modulation, the predicted frequencies will be

$$k_1 f_m \pm k_2 N_i \quad k_1 = 1, 2, 3... \quad k_2 = 1, 2, 3... \quad i = 1 \text{ or } 2 \qquad (13.10)$$

As in the case of bearing signals, the measured signal is the output of a system H to the excitation forces. The same frequencies are obviously predicted, but all component amplitudes are weighed by $H(f)$, the transfer function.

14

Delays and Echoes

Situations involving wave propagations can be encountered in systems involving flow phenomena, acoustics, vibrations, etc. Delays, dependent on the speed of propagation and geometric parameters, can be encountered, while reflections from boundaries can also create echoes. This chapter and the relevant exercises briefly introduce some approaches which are encountered in the disciplines where propagation effects need to be addressed.

14.1 System with Pure Delays

An ideal delaying system (see also Section 5.5), delaying by τ and excited by $x(t)$, will have a response $y(t)$ of

$$y(t) = Ax(t - \tau) \tag{14.1a}$$

where A is a static gain. In the frequency domain

$$Y(\omega) = X(\omega)\exp(-j\omega\tau) \tag{14.1b}$$

The impulse response and the FRF of the system are then

$$h(t) = A\delta(t - \tau)$$

$$H(\omega) = A\exp(-j\omega\tau) \tag{14.2}$$

with a phase shift proportional to the frequency. When either H or h are determined experimentally, the other can be computed via Fourier transform relations. For nonideal delay systems, the constant gain and linear frequency property may be limited to a partial frequency range. Only those signal components in this range will be shifted (without attenuation) by a constant delay.

As a special case, let us investigate a system with multiple reflections, such that the path delays are integer multiples of a single delay. The impulse response of the system is then a superposition of delayed impulses

$$\delta(t) = \sum_q \delta(t - q\tau) \tag{14.3a}$$

Discover Signal Processing: An Interactive Guide for Engineers   S. Braun
© 2008 John Wiley & Sons, Ltd

The effect of delays in the frequency domain is seen as interference in the spectra, caused by reflections. Assuming, for the sake of simplicity, a single reflection superimposed on the directly propagating signal, with a PSD of S, the resultant transfer function H_r would be

$$H_r = H[1+\exp(j2\pi f\tau)]$$

and the resultant PSD

$$S_r = S|H_r|^2 = S|H_0|^2\left|[1+\exp(-j2\pi f\tau)]\right|^2 \tag{14.3b}$$

The modul of the complex expression is zero for $f = 1/2\tau$ and maximum for $f = 1/\tau$ (as well as their integer multiples), hence the interference in the frequency domain. It is obvious that the same pattern of interference, separated by integer multiples of $1/t$, will occur for the case corresponding to Equation (14.2a).

14.2 Correlation Functions

For systems excited by random signals, correlation functions may be effective in identifying delays, as would occur in multiple path propagation situations (Bendat and Piersol, 2000). Assuming a multiple path system, where each path introduces a different delay, we model this as a single input/multiple output (SIMO) system:

$$y = \sum_q A_q x(t-\tau_q) \tag{14.4}$$

The autocorrelation of x, and the cross-correlation between x and y, are then based on expectations $E[*]$:

$$R_{xx} = E\left[x(t)x(t+\tau)\right]$$

$$R_{xy}(\tau) = E\left[x(t)y(t+\tau)\right] = E\left[x(t)\sum_q A_q x(t+\tau-\tau_q)\right] = \sum_q A_q R_{xx}(\tau-\tau_q) \tag{14.5}$$

For reasonably wideband signals, R_{xx} peaks around zero (but see further on), and the cross-correlations will show multiple peaks at locations τ_q.

We recall Exercise 7.9. The width of the autocorrelation function (at zero delay) is a function of the bandwidth of the signal. Hence this width needs to be less than the delay in order to identify them. The width of the autocorrelation is inversely proportional to the frequency bandwidth of the signal (being the IFT of the PSD, see Section 7.9). The identification of delays via the cross-correlation can become problematic for narrow band signals. For multiple delays, we need

$$|\Delta\tau| \le \frac{1}{BW} \tag{14.6}$$

where $\Delta\tau$ is the delay difference, and BW the bandwidth.

14.3 Cepstral Analysis (Braun, 1986; Randall, 2002)

The original idea followed the observation that when a delayed echo is added to a transient signal, then interferences would occur in the spectral domain. For an echo with an amplitude a_0 and delayed by t, we have specifically

$$x(t) = s(t)+a_0 s(t-\tau)$$
$$X(\omega) = S(\omega)[1+a_0^2+2a_0\cos(\omega\tau)]$$
$$|X(\omega)|^2 = |S(\omega)|^2[1+a_0^2+2a_0\cos(w\tau)] \tag{14.7}$$

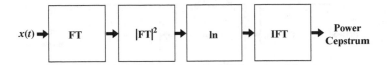

Figure T14.1

The product can be transformed to an addition by applying a log operation

$$\ln|X(\omega)|^2 = \ln|S(\omega)|^2 + \ln\left[1 + a_0^2 + 2a_0\cos(\omega\tau)\right] \tag{14.8}$$

The log being periodic, the second term of Equation (14.7) can be represented by a Fourier series with components having frequencies of $k\omega\tau$, $k = 1, 2\ldots$ Reverting to the time domain via an inverse Fourier transform (see Figure T14.1) would show spikes (δ functions) at locations superimposed on the original transient. The existence and location detection of the echo can be based on identifying these spikes, the location being obviously equal to $k\tau$.

Figure T14.2 shows an example of an echo delayed by 0.4 sec. Spikes occur at 0.8 sec (multiple of 0.8 sec). The symmetries induced by applying the FFTF, as discussed in Chapter 3, can also be noted.

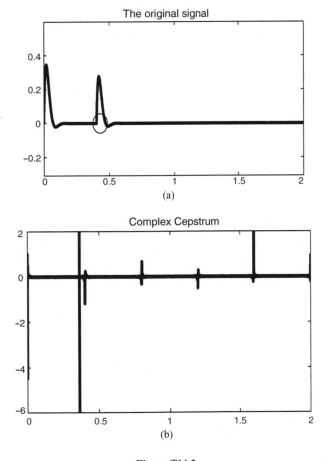

Figure T14.2

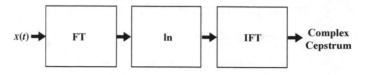

Figure T14.3

The spikes at $1.2 = 2 - 0.8$ sec and $1.6 = 2 - 0.8$ sec correspond to the negative times -0.4 and -0.8 sec.

The scheme described by Figure T14.1 is known as 'power Cepstrum' analysis. To recover the original signal, a different scheme known as 'complex Cepstrum' is used (see Figure T14.3). Comparing with the power Cepstrum, the left part shows that the absolute square operation shown in Figure T14.1 is avoided, and a complex logarithm operation is used. The right part then shows an inverse operation, where the original composite signal can be recovered. Filtering out (removing) the spikes due to the echo, would then remove the echoes, and the inverse operation would recover the original signal only.

A somewhat strange terminology is used in Cepstrum analysis – Cepstrum replacing 'spectrum', the filtering operation denoted as 'liftering'. Many other 'distorted' names used are based on similar similarities.

References

The number of available books dealing with signal processing is overwhelmingly large, and growing constantly. The relatively reduced list below reflects mainly the author's choice and familiarity, and excellent additional ones are readily available. The website of Mathworks, http://www.mathworks.com, has a list of books on signal processing with Matlab application, and might be a convenient source for additional searches.

Some of these references excel in their general didactic approaches: Ambardar (2007), Cooper and McGillen (1971), Haykin and Van Veen (1999), Porat (1992); or when exposing some specific topics, like Fourier series and modulation (Lathi, 1968) or difference equations (Gabel and Roberts, 1987). Seminal books, withstanding the passage of time (or newer versions by the same authors) are those by Bendat and Piersol (1993, 2000), Blackman and Tukey (1958), Davenport and Root (1958), Jenkins and Watts (1969), Oppenheim et al. (1999) and Papoulis (1977). Those by Candy (2005), Hayes (1996) and Stoica and Moses (2002) were chosen for their exposition of parametric methods. Some were included as enabling a simplified, albeit convenient first approach (Lyons, 2004; McMahon, 2006), and specifically the Mathwork documentation one.

While many recent books include CDs with Matlab programs, few can be considered as stressing a multimedia approach. Mentioned are McLellan et al. addressing more basic topics than the present text, and Karrenberg (2002).

While most authors strive to address a general audience, many stress topics of importance to EEs, a few are specifically oriented to others: Braun (1975, 1980, 1986), Braun and Datner (1979) and Braun and Seth (1980) to MEs; Ljung and Glad (1994) to a very general audience; and Romberg et al. (1996) to industry. The book by Stein (2000), geared to the computer engineers, was included due its original exposition of classical topics. In this category is also the forthcoming one by Shin (2008), which to some extent could be considered as complementing this present book.

Ambardar A., *Digital Signal Processing, a Modern Introduction*, Thompson, 2007

Baher H., *Analog and Digital Signal Processing*, Chichester: John Wiley & Sons, Ltd 1990

Bendat J.S., Piersol A.G., *Engineering Applications of Correlation and Spectral Analysis*, New York: Wiley-Interscience, 1993

Bendat J.S., Piersol A.G., *Random Data: Analysis and Measurement Procedures*, New York: Wiley-Interscience, 2000

Blackman R.B., Tukey J., *The Measurement of Power Spectra*, New York: Dover, 1958

Braun S., The extraction of periodic waveforms by time domain averaging, *Acustica* 32(2), 1975, 69–77

Braun S., The signature analysis of sonic bearing vibrations, *IEEE Transactions Sonics and Ultrasonics* SU-27 (6), Nov. 1980, pp. 317–327

Braun S., *Mechanical Signature Analysis*, London: Academic Press, 1986

Braun S., Datner B., Analysis of roller/ball bearings, *ASME Transactions Journal of Design* 101(1), 1979, pp. 118–128

Braun S., Seth B., Analysis of repetitive mechanism signatures, *Journal of Sound and Vibration* 70(4), 1980, 513–522

Candy J.V., *Model Based Signal Processing*, Hoboken, NJ: Wiley Interscience, 2005

Cappelini V., Constantinides A.G., Emiliani P., *Digital Filters and their Application*, Academic Press, 1978

Cavicchi T.J., *Digital Signal Processing*, New York: John Wiley & Sons, Ltd, 2000

Cooper G.H., McGillen C.D *Probabilistic Methods of Signal and System Analysis*, New York: Holt, Rinehart & Winston, 1971

Dally J.W., Riley W.F., McConnel K.G., *Instrumentation for Engineering Measurements*, New York: John Wiley & Sons, Ltd, 1993

Davenport W.B., Root W.L., *An Introduction to the Theory of Random Signals*, New York: McGraw-Hill, 1958

Doebelin E.O., *Measurement Systems: Application and Design*, Boston, Mass.: McGraw-Hill Higher Education, 2003

Feldman M., Hilbert transform, in *Encyclopedia of Vibrations*, Academic Press, 2002

Gabel R.A., Roberts R.A., *Signals and Linear Systems*, New York: John Wiley & Sons, Ltd, 1987

Hamming R.W., *Digital Filters*, Englewood Cliffs, NJ: Prentice-Hall, 1989

Hayes M., *Statistical Digital Signal Processing and Modeling*, John Wiley & Sons, Ltd, 1996

Haykin S.,Van Veen B., *Signal and Systems*, New York: John Wiley & Sons, Ltd, 1999

Jenkins G.M., Watts D.G., *Spectral Analysis and its Application*, San Francisco: Holden-Day, 1969

Karrenberg U., *An Interactive Multimedia Introduction to Signal Processing*, Springer, 2002

Lathi, B.P., *Communication Systems*, New York: John Wiley & Sons, Ltd, 1968

Ljung L., Glad T., *Modeling of dynamic systems*, Englewood Cliffs, NJ: PTR Prentice Hall, 1994

Lynn P., Fuerst W., *Introductory Signal Processing*, New York: John Wiley & Sons, Ltd, 1998

Lyons R.G., *Understanding Signal Processing*, Upper Saddle River, NJ: Prentice Hall, 2004

McClellan J.H., Shafer R.W.,Yoden M.A., *DSP first, a Multimedia Approach*, Upper Saddle River, NJ: Prentice Hall, 1998

McFadden P.D., Smith J.D., Model for the vibration produced by a single point defect in a rolling element bearing, *Journal of Sound and Vibration* 96(1), 1984, pp. 69–82

McMahon D., *Signal and systems de Mystified*, McGraw-Hill, 2006

Mathwork Documentation Signal Processing Toolbox

Naida, P.S. *Modern Digital Signal Processing*, Oxford: Alpha Science, 2006

Oehlmann H., Brie D., Tomezak M., Richard A., A method for analyzing gearbox faults using time frequency representation, *Mechanical Systems and Signal Processing* 11(4), 1997, pp. 529–545

Oppenheim A.V., Shafer R.W., Buck J.R, *Discrete Time Signal Processing*, London: Prentice Hall, 1999

Papoulis A., *Signal Analysis*, New York: McGraw-Hill, 1977

Porat B., *A Course in Digital Signal Processing*, John Wiley & Sons, Ltd, 1992

Prashad H., Ghosh M., Biswas S., Diagnostic monitoring of rolling element bearing by high–frequency resonance technique, *ASLE Transaction* 28(4), 1985, pp. 439–448

Proakis J.G., Manolakis P., *Digital Signal Processing*. Upper Saddle River, NJ: Prentice Hall, 2007

Qian S., *Joint Time-frequency Analysis: Methods and Applications*, Upper Saddle River, NJ: Prentice-Hall PTR, 1996

Randall R.B., A new method of modeling gear faults, *ASME Journal of Mechanical Design,* 104(2), 1982, pp. 259–267

Randall R.B., Cepstrum analysis method, in *Encyclopedia of Vibrations*, Academic Press, 2002

Romberg T.M., Black J.L., Ledwidge T.J., *Signal Processing for Industrial Diagnostics*. New York: John Wiley & Sons, Ltd., 1996

Root D., *An Introduction to the Theory of Random Signals and Noise,* New York: *McGraw-Hill, 1958*

Schilling R.J., Harris S.L., *Fundamentals of Digital Signal Processing*, Thompson, 2005

Schwartz M., Shaw L., *Signal Processing: Discrete Spectral Analysis, Detection and Estimation*, New York: McGraw-Hill, 1975

Shin K., Hammond J.K., Fundamentals of Signal Processing for Sound and Vibration Engineers, in preparation, to be published by John Wiley & Sons, Ltd in 2008

Stearns S.D., *Digital Signal Processing*, New York: CRC Press, 2002

Stein J., *Digital Signal Processing: a Computer Science Perspective*, John Wiley & Sons, Ltd, 2000

Stoica P., Moses R.L., *Spectral Analysis of Signals*, Prentice Hall, 2002

Wang W.J., McFadden P.D., Decomposition of gear motion signals and its application to gearbox diagnostics, *Transactions ASME Journal of Vibration and Acoustics* 117, July 1995, 363–367

White G., amplitude demodulation–a new tool for predictive maintenance, *Journal of Sound and Vibration* Sept 1991, pp. 14–19

White P., Time frequency methods, in *Encyclopedia of Vibrations*, Academic Press, 2002

Wu S., Tobin T.H., Chow C., Signature analysis for mechanical systems via the DDS monitoring technique, *ASME Transactions Journal of Mechanical design* 102(2), 1980, 217–221

Index